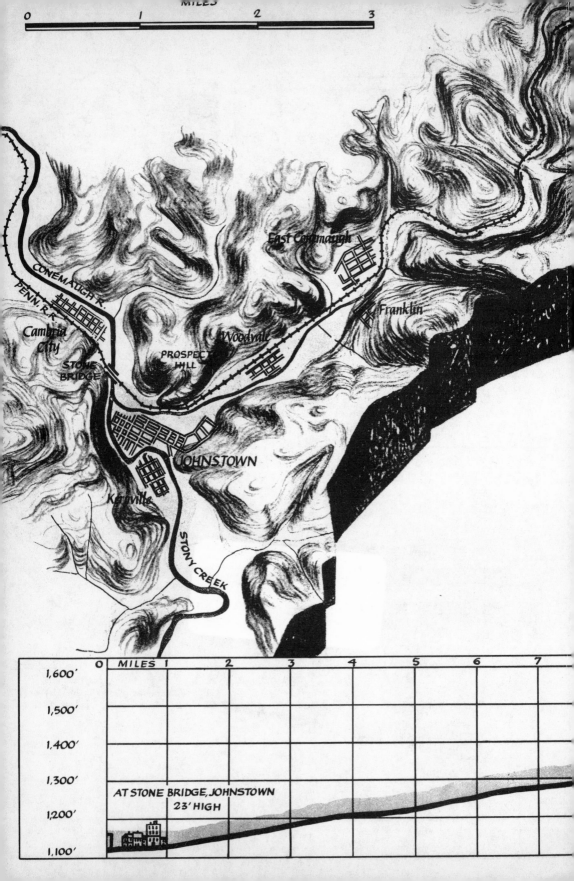

MILES

0   1   2   3

East Conemaugh

CONEMAUGH R.

PENN. R.R.

Franklin

Cambria City

Woodvale

PROSPECT HILL

STONE BRIDGE

JOHNSTOWN

Kernville

STONY CREEK

	0	MILES 1	2	3	4	5	6	7
1,600'								
1,500'								
1,400'								
1,300'								
1,200'	AT STONE BRIDGE, JOHNSTOWN 23' HIGH							
1,100'								

THE PATH OF THE FLOOD

palacios

ALSO BY DAVID MCCULLOUGH

The Great Bridge
The Path Between the Seas
Mornings on Horseback
Truman
Brave Companions
John Adams

THE
JOHNSTOWN
FLOOD

by
David McCullough

Simon & Schuster Paperbacks
New York   London   Toronto   Sydney

Simon & Schuster Paperbacks
Rockefeller Center
1230 Avenue of the Americas
New York, NY 10020

Copyright © 1968 by David McCullough
All rights reserved,
including the right of reproduction
in whole or in part in any form.

SIMON & SCHUSTER PAPERBACKS and colophon are registered trademarks
of Simon & Schuster, Inc.

For information regarding special discounts for bulk purchases,
please contact Simon & Schuster Special Sales at
1-800-456-6798 or business@simonandschuster.com

Manufactured in the United States of America

30   29   28   27

The Library of Congress has cataloged the Touchstone edition as follows:
McCullough, David G.
    The Johnstown flood.
    (A Touchstone book)
    Bibliography: p.
    1. Floods—Pennsylvania—Johnstown.   2. Johnstown (Pa.)
—History.   I. Title.
[F159.J7M16   1987]   974.8'77   86-26056

ISBN 0-671-20714-8

For Rosalee

PENNSYLVANIA

AREA
OF MAP

Pittsburgh

Johnstown

PENN. R.R.

Altoona

Ebensburg

Cresson

Hollidaysburg

Nineveh
(now Seward)

LITTLE CONEMAUGH R.

Lilly

Bolivar

Sang
Hollow

South Fork

New Florence

LAKE
CONEMAUGH

JOHNSTOWN

LAUREL HILL

STONY CR.

ALLEGHENY MOUNTAINS

Stoystown

Bedford

B. & O. R.R.

Somerset

palacios

Acknowledgments

The material for this book was gathered from the files of newspapers, from unpublished reminiscences, from letters and diaries, from Johnstown Flood "histories" that were best sellers in their day and from books and pamphlets that were privately printed, from court records, engineering reports, local histories, and rare old maps, from old photographs, and from hours of taped conversations with survivors of the Johnstown Flood.

A bibliography is included at the back of the book, but I want to acknowledge my indebtedness to four works in particular: J. J. McLaurin's *The Story of Johnstown*, which for all its Victorian embellishments and inaccuracies is the finest by far of the books "gotten up" by journalists in 1889; The Reverend David J. Beale's *Through the Johnstown Flood*, in many ways the best book on the flood and unquestionably the best-written and most reliable of accounts by survivors; *A History of Johnstown and the Great Flood of 1889: A Study of Disaster and Rehabilitation*, which is a doctoral thesis written by the late Nathan D. Shappee and the only scholarly study of the disaster; and a recently discovered transcription of testimony taken by the Pennsylvania Railroad during the summer of 1889, which has been invaluable. (Most of the dialogue in Chapters 3 and 4, for example, has been taken directly from this transcription, which, in all, runs to nearly 500 typewritten pages, and no part of which has been previously published.)

I very gratefully acknowledge my debt to the following flood survivors who kindly gave me so much of their time to talk about their experiences, some of whom have since passed on: Mrs. Kate Miltenberger, Mr. and Mrs. Harry Hesselbein, and David Fetterman, all of Johnstown; U. Ed Schwartzentruver of South Fork; Mrs. Gertrude

11

Quinn Slattery of Wilkes-Barre, who also read the manuscript; and Dr. Victor Heiser of New York, who, in addition to vivid descriptions of his own experiences, supplied wonderful insights into the Johnstown of the 1880's, and who read the manuscript.

I am grateful also to the two Johnstown ladies, both survivors, who shared memories of their illustrious family, but who asked that I not mention their names.

I wish to thank too the many others in Johnstown who were helpful, and especially the following: Irving London, who led me to the Pennsylvania Railroad testimony and who made available his extraordinary collection of flood photographs; Harold Strayer; Gustaf Hultman of the National Park Service; Elit Felix; Walter Krebs, president of the Johnstown *Tribune-Democrat*, who made available the paper's reference library and files; Don Matthews, Jr., also of the *Tribune-Democrat*, who made numerous helpful suggestions; Frank Dell and Ron Stephenson of Station WJAC; the City Clerk's office; and the staff of the Cambria Public Library.

In addition, I am much indebted to Edna Lehman, who made available important documentary material at the Cambria County Historical Society at Ebensburg; to the late Robert Heppenstall of Pittsburgh, for information on his father's heroic action; to Mrs. John E. Hannon, Sr., of Detroit, who wrote to me at length about her grandfather, W. Horace Rose; to Mrs. Bernard McGuire of Cresson, who let me borrow a diary kept by her grandfather, Isador Lilly of Ebensburg; to Dr. Philip Bishop, Dr. John White, and Donald Berkebile of the Smithsonian Institution; to Mr. and Mrs. Fred Livengood, Sr., of Somerset; to the late Mrs. O. C. Gaub of Pittsburgh; to Evan Stineman of South Fork; and to Al Danel of St. Michael, Pennsylvania.

Also I wish to express my gratitude for the facilities offered by the staff of the Pennsylvania Room at the Carnegie Library, Pittsburgh; the Allegheny County Court of Claims, Pittsburgh; the Local History and Genealogy Room at the New York Public Library; the New York Historical Society; the Library of Congress; the University of Pennsylvania; the University of Pittsburgh; the Yale University Library; the White Plains Public Library; and the Boston Public Library.

And for their suggestions and encouragement I wish finally to thank Walter McQuade, Walter Lord, Roger Butterfield, David Plowden, Heywood Broun, Jr., David Allison, Frank Fogarty, James Morrison, Royall O'Brien, Charles T. Siebert, Jr.; Audre Proctor, who typed the manuscript; my mother and father; and my wife, Rosalee.

D. McC.

Contents

List of Illustrations

THE
JOHNSTOWN
FLOOD

"We are creatures of the moment; we live from one little space to another; and only one interest at a time fills these."

—*William Dean Howells*
in A Hazard of New Fortunes, *1889.*

I

The sky was red

-1-

Again that morning there had been a bright frost in the hollow below the dam, and the sun was not up long before storm clouds rolled in from the southeast.

By late afternoon a sharp, gusty wind was blowing down from the mountains, flattening the long grass along the lakeshore and kicking up tiny whitecaps out in the center of the lake. The big oaks and giant hemlocks, the hickories and black birch and sugar maples that crowded the hillside behind the summer colony began tossing back and forth, creaking and groaning. Broken branches and young leaves whipped through the air, and at the immense frame clubhouse that stood at the water's edge, halfway among the cottages, blue wood smoke trailed from great brick chimneys and vanished in fast swirls, almost as though the whole building, like a splendid yellow ark, were under steam, heading into the wind.

The colony was known as the South Fork Fishing and Hunting Club. It was a private summer resort located on the western shore of a mountain lake in Cambria County, Pennsylvania, about halfway between the crest of the Allegheny range and the city of

19

Johnstown. On the afternoon of Thursday, May 30, Memorial Day, 1889, the club was not quite ten years old, but with its gaily painted buildings, its neat lawns and well-tended flower beds, it looked spanking new and, in the gray, stormy half-light, slightly out of season.

In three weeks, when the summer season was to start, something like 200 guests were expected. Now the place looked practically deserted. The only people about were a few employees who lived at the clubhouse and some half dozen members who had come up from Pittsburgh for the holiday. D. W. C. Bidwell was there; so were the young Clarke brothers, J. J. Lawrence, and several of the Sheas and Irwins. Every now and then a cottage door slammed, voices called back and forth from the boathouses. Then there would be silence again, except for the sound of the wind.

Sometime not long after dark, it may have been about eight thirty, a young man stepped out onto the long front porch at the clubhouse and walked to the railing to take a look at the weather. His name was John G. Parke, Jr. He was clean-shaven, slight of build, and rather aristocratic-looking. He was the nephew and namesake of General John G. Parke, then superintendent of West Point. But young Parke was a rare item in his own right for that part of the country; he was a college man, having finished three years of civil engineering at the University of Pennsylvania. For the present he was employed by the South Fork Fishing and Hunting Club as the so-called "resident engineer." He had been on the job just short of three months, seeing to general repairs, looking after the dam, and supervising a crew of some twenty Italian laborers who had been hired to install a new indoor plumbing system, and who were now camped out of sight, back in the woods.

In the pitch dark he could hardly see a thing, so he stepped down the porch stairs and went a short distance along the boardwalk that led through the trees to the cottages. The walk, he noticed, was slightly damp. Apparently, a fine rain had fallen sometime while he was inside having his supper. He also noticed that though the wind was still up, the sky overhead was not so dark as before; indeed, it seemed to be clearing off some. This was not what he had expected. Windstorms on the mountain nearly always meant a heavy downpour almost immediately after—"thunder-gusts" the local men called them. Parke had been through several already in the time he had been at the lake and knew what to expect.

It would be as though the whole sky were laying siege to the burly landscape. The rain would drum down like an unyielding river. Lightning would flash blue-white, again and again across the sky, and thunderclaps would boom back and forth down the valley like a cannonade, rattling every window along the lakeshore.

Then, almost as suddenly as it had started, the siege would lift, and silent, milky steam would rise from the surface of the water and the rank smell of the sodden forest floor would hang on in the air for hours.

Tonight, however, it appeared there was to be no storm. Parke turned and walked back inside. About nine-thirty he went upstairs, climbed into bed, and went to sleep.

About an hour and a half later, very near eleven, the rain began. It came slamming through the blackness in huge wind-driven sheets, beating against the clubhouse, the tossing trees, the lake, and the dark, untamed country that stretched off in every direction for miles and miles.

The storm had started out of Kansas and Nebraska, two days before, on May 28. The following day there had been hard rains in Kansas, Missouri, Illinois, Michigan, Indiana, Kentucky, and Tennessee. Trains had been delayed, roads washed out. In Kansas, along the Cottonwood River, a dozen farms had been flattened by tornado-force winds and several people had been killed. In northern Michigan and parts of Indiana there had been sudden snow squalls. Warnings had been telegraphed east. On the night of the 29th the U.S. Signal Service issued notices that the Middle Atlantic states were in for severe local storms. On the morning of May 30 all stations in the area reported "threatening weather."

When the storm struck western Pennsylvania it was the worst downpour that had ever been recorded for that section of the country. The Signal Service called it the most extensive rainfall of the century for so large an area and estimated that from six to eight inches of rain fell in twenty-four hours over nearly the entire central section. On the mountains there were places where the fall was ten inches.

But, at the same time, there were astonishing disparities between the amount of rainfall at places within less than a hundred-mile radius. At the South Fork Fishing and Hunting Club, for example, a pail left outside overnight would have five inches of water in it the next morning when the rain was still coming down. The total rainfall at the clubhouse would be somewhere near seven inches. In Pittsburgh, just sixty-five miles to the west as the crow flies, the total rainfall would be only one and a half inches.

But as the storm beat down on the mountain that night, John G. Parke, Jr., who would turn twenty-three in less than a month, slept on, never hearing a thing.

-2-

Most of the holiday crowds were back from the cemetery by the time the rain began Thursday afternoon. It had been the customary sort of Memorial Day in Johnstown, despite the weather.

People had been gathering along Main Street since noon. With the stores closed until six, with school out, and the men off from the mills, it looked as though the whole town was turning out. Visitors were everywhere, in by special trains from Somerset, Altoona, and other neighboring towns. The Ancient Order of Hibernians, "a stalwart, vigorous looking body of men," as the Johnstown *Tribune* described them, was stopping over for its annual state convention. Hotels were full and the forty-odd saloons in Johnstown proper were doing a brisk business.

The Reverend H. L. Chapman, who lived two doors off Main, in the new Methodist parsonage facing the park, later wrote, "The morning was delightful, the city was in its gayest mood, with flags, banners and flowers everywhere . . . we could see almost everything of interest from our porch. The streets were more crowded than we had ever seen before."

The parade, late starting as always, got under way about two-thirty, marched up Main, past the Morrell place, on by the Presbyterian Church and the park, clear to Bedford Street. There it turned south and headed out along the river to Sandy Vale, where the war dead were buried. The fire department marched, the Morrellville Odd Fellows, the Austrian Music Society, the Hornerstown Drum Corps, the Grand Army Veterans, and the Sons of

Veterans, and half a dozen or more other groups of various shapes and sizes, every one of them getting a big cheer, and especially the Grand Army men, several of whom were beginning to look as though the three-mile tramp was a little more than they were up to.

How much things had changed since they had marched off to save the Union! It had been nearly thirty years since Lincoln had first called for volunteers. Grant and Lee were both dead, and there were strapping steelworkers with thick, black mustaches standing among the crowds along Main Street who had been born since Appomattox.

At the start of the war Johnstown had been no more than a third the size it was now; and ten years before that, it had been nothing but a sleepy little canal town with elderbushes growing high along Main, and so quiet you could hear the boat horns before the barges cleared the bend below town.

But ever since the war, with the west opening up, the Cambria Iron Company had had its giant three-ton converters going night and day making steel for rails and barbed wire, plowshares, track bolts, and spring teeth for harrows. The valley was full of smoke, and the city clanked and whistled and rumbled loud enough to be heard from miles off. At night the sky gleamed so red it looked as though the whole valley were on fire. James Quinn, one of Johnstown's most distinguished-looking Grand Army veterans and its leading dry-goods merchant, enjoyed few sights more. "The sure sign of prosperity," he called it.

Years after, Charlie Schwab, the most flamboyant of Carnegie's men, described the view of Johnstown from his boyhood home in the mountain town of Loretto, nearly twenty miles to the northeast.

"Along toward dusk tongues of flame would shoot up in the pall around Johnstown. When some furnace door was opened the evening turned red. A boy watching from the rim of hills had a vast arena before him, a place of vague forms, great labors, and dancing fires. And the murk always present, the smell of the foundry. It gets into your hair, your clothes, even your blood."

Most of the men watching the parade that Memorial Day would have taken a somewhat less romantic view. In the rolling mills they worked under intense heat on slippery iron floors where molten metal went tearing by and one false step or slow reaction could mean horrible accidents. Most of them worked a ten- or even twelve-hour day, six days a week, and many weeks they worked the hated "long turn," which meant all day Sunday and on into Monday. If they got ten dollars for a week's work they were doing well.

A visiting journalist in 1885 described Johnstown as "new, rough, and busy, with the rush of huge mills and factories and the throb of perpetually passing trains." The mills were set just below town in the gap in the mountains where the Conemaugh River flows westward. On the hillside close to the mills the trees had turned an evil-looking black and grew no leaves.

Johnstown of 1889 was not a pretty place. But the land around it was magnificent. From Main Street, a man standing among the holiday crowds could see green hills, small mountains, really, hunching in close on every side, dwarfing the tops of the houses and smokestacks.

The city was built on a nearly level flood plain at the confluence of two rivers, down at the bottom of an enormous hole in the Alleghenies. A visitor from the Middle West once commented, "Your sun rises at ten and sets at two," and it was not too great an exaggeration.

The rivers, except in spring, appeared to be of little consequence. The Little Conemaugh and Stony Creek, or the Stony Creek, as everyone in Johnstown has always said (since it is *the* Stony Creek *River*), are both more like rocky, oversized mountain streams than rivers. They are about sixty to eighty yards wide. Normally their current is very fast; in spring they run wild. But on toward August, as one writer of the 1880's said, there are places on either river where a good jumper could cross on dry stones.

The Little Conemaugh, which is much the swifter of the two, rushes in from the east, from the Allegheny Mountain. It begins near the very top of the mountain, about eighteen miles from Johnstown, at a coal town called Lilly. Its sources are Bear Rock Run and Bear Creek, Trout Run, Bens Creek, Laurel Run, South Fork Creek, Clapboard Run and Saltlick Creek. From an elevation of 2,300 feet at Lilly, the Little Conemaugh drops 1,147 feet to Johnstown.

The Stony Creek flows in from the south. It is a broader, deeper river than the other and is fed by streams with names like Beaver Dam Run, Fallen Timber Run, Shade Creek, and Paint Creek. Its total drainage is considerably more than that of the Little Cone-maugh, and until 1889 it had always been thought to be the more dangerous of the two.

When they meet at Johnstown, the rivers form the Cone-maugh, which, farther west, joins the Loyalhanna to form the Kis-kiminetas, which in turn flows into the Allegheny about eighteen miles above Pittsburgh.

At Johnstown it was as though the bottom had dropped out of the old earth and left it angry and smoldering, while all around, the long, densely forested ridges, "hogbacks" they were called, rolled off in every direction like a turbulent green sea. The climb up out of the city took the breath right out of you. But on top it was as though you had entered another world, clean, open, and sweet-smelling.

In 1889 there were still black bear and wildcats on Laurel Hill to the west of town. Though the loggers had long since stripped the near hills, there were still places within an hour's walk from Main Street where the forest was not much different than it had been a hundred years before.

Now and then an eagle could still be spotted high overhead. There were pheasants, ruffed grouse, geese, loons, and wild turkeys that weighed as much as twenty pounds. Plenty of men marching in the parade could remember the time before the war when there had been panthers in the mountains big enough to carry off a whole sheep. And it had been only a few years earlier when passenger pigeons came across the valley in numbers beyond belief. One Jan-uary the *Tribune* wrote: "On Saturday there were immense flocks of wild pigeons flying over town, but yesterday it seemed as if all the birds of this kind at present in existence throughout the entire country were engaged in gyrating around overhead. One flock was declared to be at least three miles in length by half a mile wide."

Still, many days there were in the valley itself when the wind swept away the smoke and the acrid smell of the mills and the air was as good as a man could ask for. Many nights, and especially in winter, were the way mountain nights were meant to be, with mil-lions of big stars hanging overhead in a sky the color of coal.

Looking back, most of the people who would remember

Johnstown as it was on that Memorial Day claimed it was not as unpleasant a place as one might imagine. "People were poor, very poor by later standards," one man said, "but they didn't know it." And there was an energy, a vitality to life that they would miss in later years.

Many of the millworkers lived in cheap, pine-board company houses along the riverbanks, where, as the *Tribune* put it, "Loud and pestiferous stinks prevail." But there were no hideous slums, such as had spread across the Lower East Side of New York or in Chicago and Pittsburgh. The kind of appalling conditions that would be described the next year by Jacob Riis in his *How the Other Half Lives* did not exist then in Johnstown. No one went hungry, or begging, though there were always tramps about, drifters, who came with the railroad, heading west nearly always, knocking at back doors for something to eat.

They were part of the landscape and people took them for granted, except when they started coming through in big numbers and there were alarming stories in the papers about crowds of them hanging around the depot.

One diary, kept by a man who lived outside of town, includes a day-by-day tramp count. "Wednesday, May 1, 1889, Two Tramps . . . Thursday, May 2, Two Tramps," and so on, with nearly a tramp or two every day, week after week.

New people came to town, found a job or, if not, moved on again, toward Pittsburgh. But for most everyone who decided to stay there was work. Although lately, Johnstown men, too, had been picking up and going west to try their luck at the mills in St. Louis or the mines in Colorado. And lately the jobs they left behind were being filled by "hunkies" brought in to "work cheap."

The idea did not please people much. Nor did it matter whether the contract workers were Italians, Poles, Hungarians, Russians, or Swedes; they were all called Hungarians, "bohunks" or "hunkies." But so far, and again unlike the big cities, Johnstown had only a few such men, and most of them lived in Cambria City, just down the river, beyond the new stone bridge that carried the main line of the Pennsylvania across the Conemaugh.

The vast majority of the people who lined Main Street watching the parade were either Irish, Scotch-Irish, or Cornish (Cousin Jacks, they were called), German or Welsh, with the Germans and the Welsh greatly outnumbering all the rest. There were some Negroes, but not many, and a few of the leading merchants were Jews.

The Germans and the Welsh had been the first settlers. More of them, plus the Scotch-Irish, had come along soon after to work in the mines and first forges. Quite a few of the big Irishmen in the crowd had come in originally to build the railroad, then stayed on. Johnstown had been an active stop along the Underground Railroad, and a few of the Negroes had come in that way. Others of them came later to work in the tannery.

There were German and Welsh churches in town, a German newspaper, and several Irish fortunetellers. Welsh and German were spoken everywhere, along with enough other brogues, burrs, and twangs to make a "plain American" feel he was in a country of "feriners," or so it often seemed.

The first white settlers in the valley had been Solomon and Samuel Adams and their sister Rachael, who came over the Allegheny Mountains from Bedford about 1771 and cleared a patch of land near the Stony Creek. Until then the place had been known as Conemack Old Town, after a Delaware Indian village that stood about where the Memorial Day parade had gathered that noon at the foot of Main.

Samuel Adams and an Indian killed each other in a knife fight, and the traditional story is that Rachael was also killed by Indians soon after. Solomon made a fast retreat back to the stockade at Bedford, and it was not for another twenty years or thereabouts that the first permanent settler arrived. In 1794, about the time President Washington was sending an army over the mountains to put down the so-called Whiskey Rebellion in Pittsburgh, Joseph Schantz, or Johns, an Amish farmer from Switzerland, came into the valley with his wife and four children. He cleared off thirty acres between the rivers, raised a cabin, planted an orchard, and laid out a village which he called Conemaugh Old Town—or just Conemaugh—and which he had every hope for becoming the county seat.

When the county was established in 1804 and given the old Latin name for Wales—Cambria—Ebensburg, a mountain village fifteen miles to the north, was picked as county seat. Three years later Joseph Johns sold his village and moved on.

The next proprietor was a long-haired "York County Dutchman" (a Pennsylvania German) named Peter Levergood, and from then until the canal came through, the town remained no more than a backwoods trading center. But with the arrival of the canal it became the busiest place in the county. By 1835 Johnstown, as it was by then known, had a drugstore, a newspaper, a Presbyterian church, and a distillery. By 1840 its population, if the nearby settlements were counted, had probably passed 3,000. Then, in the 1850's, the Pennsylvania Railroad came through, the Cambria Iron Company was established, and everything changed.

By the start of the 1880's Johnstown and its neighboring boroughs had a total population of about 15,000. Within the next nine years the population doubled. On the afternoon of May 30, 1889, there were nearly 30,000 people living in the valley.

Properly speaking Johnstown was only one of several boroughs—East Conemaugh, Woodvale, Conemaugh, Cambria City, Prospect, Millville, Morrellville, Grubbtown, Moxham, Johnstown —which were clustered between the hills, packed in so tight that there was scarcely room to build anything more.

Petty political jealousies and differences over taxes had kept them from uniting. As it was there was no telling where one began or the other ended unless you knew, which, of course, everyone who lived there did. Millville, Prospect, and Cambria City, it was said, lived on the pay roll of the Cambria mills; Conemaugh lived on the Gautier wire works, Woodvale on the woolen mills there, and Johnstown, in turn, lived on all the rest of them. Johnstown was the center of the lot, geographically and in every other way. It was far and away the largest, with a population of its own of perhaps 10,000 by 1889, which was four times greater than even the biggest of the others. The banks were there, the hotels, the jail, and a full-time police force of nine.

There were five-story office buildings on Main and up-to-date stores. The town had an opera house, a night school, a library, a remarkable number of churches, and several large, handsome houses, most of which were owned by men high up in the Iron Company.

Much would be written later on how the wealthy men of Johnstown lived on the high ground, while the poor were crowded into the lowlands. The fact was that the most imposing houses in town were all on Main Street, and one of the largest clusters of company houses was up on Prospect Hill.

The rest of the people lived in two- and three-story frame houses which, often as not, had a small porch in front and a yard with shade trees and a few outbuildings in back. Nearly everyone had a picket fence around his property, and in spite of its frenzied growth, the city still had more than a few signs of its recent village past.

On the 22nd of May, for example, the town fathers had gathered at the City Council chambers to settle various matters of the moment, the most pressing of which was to amend Section 12 of Chapter XVI of the Codified Ordinance of the Borough of Johnstown. The word "cow" was to be inserted after "goat" in the third line, so that it would from then on read: "Section 12. Any person who shall willfully suffer his horse, mare, gelding, mule, hog, goat, *cow*, or geese to run at large within the Borough shall for each offense forfeit and pay for each of said animals so running at large the sum of one dollar . . ."

Life was comparatively simple, pleasures few. There were Saturday night band concerts in the park, and lectures at the library. Sundays half the town put on its best and went walking. Families would pick one of the neighboring boroughs and walk out and back, seeing much and, naturally, being seen all along the way.

There was a new show at the Washington Street Opera House every other night or so. Thus far in 1889 it had been an especially good season, with such favorites as *Zozo the Magic Queen* (which brought its "splendid production" in "OUR OWN SPECIAL SCENERY CAR") and *Uncle Tom's Cabin* appearing in a single week. (Harriet Beecher Stowe's little drama had also changed considerably since the years before the war. The Johnstown performance, for example, featured "a pack of genuine bloodhounds; two Topsies; Two Marks, Eva and her Pony 'Prince'; African Mandolin Players; 'Tinker' the famous Trick Donkey.")

There was also the Unique Rink for roller skating, a fad which seemed to be tapering off some that spring. There was superb fishing along the Conemaugh and the Stony Creek in spring and summer. Downstream from town the river was stained by waste dumped from the mills, but above town the water still ran clear between sun-bleached boulders and was full of catfish, sunfish, mullet, walleyed pike that everyone mistakenly called salmon, trout, eels, and speedy, mud-colored crawfish.

In spring, too, there was nearly always a good brawl when the circus came to town. In the fall, when the sour gums turned blood-red against the pines, there was wonderful hunting on the mountains, and fresh deer hanging from butcher shop meathooks was one of the signs of the season. In winter there were sleigh rides to Ebensburg, tobogganing parties, and ice skating at the Von Lunen pond across from the new Johnson Street Rail works up the Stony Creek at Moxham.

And year round there was a grand total of 123 saloons to choose from in the greater Johnstown area, ranging from California Tom's on Market Street to the foul-smelling holes along the back alleys of Cambria City. California Tom Davis had been a forty-niner. He was one of the colorful characters of Johnstown and the back room of his saloon was the favorite gathering place for those professional men and Cambria Iron officials who liked to take a sociable drink now and then.

But the average saloon was simply a place where a working-man could stop off at the end of the day to settle the fierce thirst the heat of the furnaces left him with, or to clear the coal dust from his throat. He was always welcome there, without a shave or a change of clothes. It was his club. He had a schooner of beer or a shot; most of the time he spent talking.

Like any steel town Johnstown had a better than average number of hard-line drinking men. On payday Saturdays the bartenders were the busiest people in town. And week after week Monday's paper carried an item or two about a "disturbance" Saturday night on Washington Street or in Cambria City, and published the names of two or three citizens who had spent the night in the lockup for behaving in "frontier fashion."

For those of still earthier appetite there was Lizzie Thompson's place on Frankstown Hill, at the end of Locust Street. It was the best-known of the sporting houses, but there were others too, close by, and on Prospect Hill. And one spring a similar enterprise had flourished for weeks in the woods outside of town, when several itinerant "soiled doves," as the *Tribune* called them, set up business in an abandoned coal mine.

But primarily, life in Johnstown meant a great deal of hard work for just about everybody. Not only because that was how life was then, but because people had the feeling they were getting somewhere. The country seemed hell-bent for a glorious new age, and Johnstown, clearly, was right up there booming along with the best of them. Pittsburgh and Chicago were a whole lot bigger, to be sure, and taking a far bigger part of the business. But that was all right. For Johnstown these were the best years ever.

Progress was being made, and it was not just something people were reading about. It was happening all around them, touching their lives.

Streets were bright at night now with sputtering white arc lights. There was a new railroad station with bright-colored awnings. The hospital was new; two new business blocks had been finished on Main. A telephone exchange had begun service that very year, in January, and already there were more than seventy phones in town. Quite a few houses had new bathrooms. The Hulbert House, the new hotel on Clinton Street, had an elevator and steam heat.

There was a street railway out to Woodvale and another up to Moxham. Almost everyone had electricity or natural gas in his home. There were typewriters in most offices, and several people had already bought one of the new Kodak "detective" cameras. "Anybody can use the Kodak," the advertisements said.

Inventions and changes were coming along so fast that it was hard to keep up with them all. The town had no debts, taxes were low, and the cost of things was coming down little by little.

Of course, there were some who looked askance at so much change and liked to talk about the old days when they said there had never been so much drinking, no prostitution, and men could still do a day's work without complaining. There was also strong resentment against the company, and quiet talk of trouble to come, though it would have been very hard for most men among those Memorial Day crowds to have imagined there ever being an actual strike in the mills. The miners had tried it, twice, and both times the company had clamped down with such speed and decisiveness that the strikes had been broken in no time.

And there were floods and fires and, worst of all, epidemics that hit so swiftly and unexpectedly, terrifying everyone, and killing so many children. The last bad time had been in 1879, when diphtheria killed 132 children within a few months. Death was always near, and there was never any telling when it would strike again.

Year in, year out men were killed in the mills, or maimed for life. Small boys playing around the railroad tracks that cut in and out of the town would jump too late or too soon and lose a leg or an arm, or lie in a coma for weeks with the whole town talking about them until they stopped breathing forever.

But had not life always been so? Was not hard work the will of the Lord? ("In the sweat of thy face shalt thou eat bread, till thou return unto the ground . . .") And yes, death too? (". . . for dust thou art, and unto dust shalt thou return.")

And besides, was it not a fine thing to be where there was so much going on, so much to keep a man busy and his family eating regularly?

So far it had been a good year. Except for measles the town seemed pretty healthy. Talk was that it would be a good summer for steel. Prices might well improve, and perhaps wages with them, and there would be no labor trouble to complicate things, as there would probably be in Pittsburgh.

The Quicksteps, Johnstown's beloved baseball team, had made a rather poor showing so far, losing to Braddock, Greensburg, and McKeesport in a row; but they had beaten Altoona once, and most people felt that about made up for it. The newspapers were full of stories about the World's Fair opening in Paris, and its Eiffel Tower, and about the Oklahoma Territory opening up out west. Towers of structural steel could reach nearly to the heavens, and Americans could turn a dusty prairie into farms and whole new cities overnight. It was some time to be alive.

But perhaps best of all there seemed such a strong spirit of national unity everywhere. The Constitutional government those Grand Army veterans had fought for had just celebrated its one hundredth birthday that spring and there had been quite a to-do about it in the newspapers and picture magazines. "A nation in its high hour of imperial power and prosperity looks back a hundred years to its obscure and doubtful beginning . . ." one article began. That the next one hundred years would be better still, bringing wondrous advantages and rewards to millions of people who had also come from "obscure and doubtful" beginnings, seemed about

as certain as anything could be; and especially in a place like Johnstown, on a day when flags were flying from one end of town to the
other, and the "Boys in Blue" were marching again.

When the rain started coming down about four o'clock, it was
very fine and gentle, little more than a cold mist. Even so, no one
welcomed it. There had already been more than a hundred days of
rain that year, and the rivers were running high as it was. The first
signs of trouble had been a heavy snow in April, which had melted
almost as soon as it came down. Then in May there had been eleven
days of rain.

The rivers ran high every spring. That was to be expected.
Some springs they ran so high they filled the lower half of town to
the top doorstep. A few times the water had been level with first-
floor windows along several streets. Floods had become part of the
season, like the dogwood blooming on the mountain. Yet, each
year, there was the hope that perhaps this time the rivers might
behave themselves.

It had already been such a curious year for weather. A tornado
in February had killed seventeen people in Pittsburgh. Not much
had happened in Johnstown, but the wind carried off a tin church
roof at Loretto. The April snow had been the heaviest of the whole
year, with fourteen inches or more in the mountains. And all
through May, temperatures had been bouncing every which way,
up in the eighties one day, down below freezing two nights later,
then back to the eighties again. Now it felt more like March than
May.

About five the rain stopped and left everything freshly rinsed
looking. The Reverend Chapman, back on his front porch after
participating at the graveside ceremonies, sat gazing at the park
with its big elms and draped chain fence, and thought to himself
that he had seldom looked upon a lovelier scene. Or at least so he
wrote later on.

The Reverend had been in Johnstown only a few years, and it
had been just the month before that he and his wife, Agnes, had
moved into the new parsonage. He had grown up along the canal to
the west of Johnstown, at Blairsville, where his father had eked out
a living painting decorative scenes and designs on packet boats. His
first church had been in Ligonier, on the other side of Laurel Hill.

Later there had been churches in New Florence and Bolivar, down the Conemaugh, and half a dozen other places between Johnstown and Pittsburgh. He liked every one of them, he said, but Johnstown was something special. His stone church next door was the largest in town, a landmark, and except for St. Joseph's, the German Catholic church over in Conemaugh borough, no church had a larger membership. His neighbors across the park, Dr. Lowman and John Fulton, were the finest sort of Christian gentlemen, and their homes were as elegantly furnished as any in town. The Dibert bank, Griffith's drugstore, the post office and the *Tribune* offices on the floor above it, were all but a few steps from the Reverend's front door. The parsonage faced on to Franklin Street, at almost the exact dead center of Johnstown.

Night settled in and the lights came on along Franklin and Main. A few blocks away William Kuhn and Daisy Horner were being married in a small ceremony at the bride's home. At the Opera House Mr. Augustin Daly's New York production of *A Night Off*, "The comedy success of two continents," was playing to a small house. Daly was the foremost theatrical producer of the day, and *A Night Off* had been his biggest hit for several years. Like some of his other productions, it was an adaptation from a German comedy, a fact which the Johnstown audience undoubtedly appreciated.

Other than that not much else was going on. Because of the holiday there had been no paper that morning, but according to Wednesday's *Tribune*, rainstorms were expected late that Thursday; tomorrow, Friday, was to be slightly warmer. The barometric pressure was reported at thirty, temperature from forty-six to sixty-five, humidity at sixty-nine per cent.

About nine the rain began again, gentle and quiet as earlier. But an hour or so later it started pouring and there seemed no end to it. "Sometime in the night," according to Chapman, "my wife asked if it were not raining very hard, and I being very sleepy, barely conscious of the extraordinary downpour simply answered, 'Yes,' and went to sleep, thinking no more of it until morning."

-3-

George Heiser's day did not end until after ten. It was his practice to keep the store open until then. With the saloons along Washington Street doing business on into the night, there were generally people about and he could pick up a little more trade. Either he or his wife Mathilde would be looking after things behind the counter, in among the queen's ware and the barrels of sugar and crackers, the cases of Ewarts tobacco and yellow laundry soap, needles, spools, pins, and Clark's "O-N-T" (Our New Thread).

George and Mathilde Heiser, and their sixteen-year-old son, Victor, lived upstairs over the store. They had been at the same location, 224 Washington, for several years now and, at long last, business was looking up. At fifty-two, for the first time in his life, George Heiser was getting on in the world.

Washington Street ran parallel to Main, two blocks to the north. From the Stony Creek over to the Little Conemaugh, the east-west streets—the "up" streets they were called—ran Vine, Lincoln, Main, Locust, Washington. Then came the B & O tracks and the B & O depot which was directly across the street from the Heiser store. Beyond the tracks were two more streets, Broad and Pearl, then the Little Conemaugh, and on the other side of that rose Prospect Hill, steep as a roof. With the town growing the way it had been, the Baltimore & Ohio had brought in a spur from Somerset eight years before to try to take away some of the freight business from the Pennsylvania. An old schoolhouse had been converted into a depot, and the steady night and day racket of the trains right by their window had made quite a difference to the Heisers. But with the store doing as well as it was, George Heiser had few complaints.

George had been troubled by bad luck much of his life. During the war, at Fredericksburg, when his unit had been making a rapid withdrawal, his companion in the ranks, a fellow named Pike, got hit and went down grabbing on to George's leg and pleading for George not to leave him. As a result they were both captured and George spent the rest of the war in Libby Prison. Then, years later, at a time when he seemed to be getting nowhere in Johnstown, George had gone off north to Oil City hoping to strike it rich. He wound up running a butcher shop instead and in no time was back in Johnstown, flat broke, wiped out by fire and his own lack of business sense.

George had not marched in the parade that afternoon. His blue uniform seldom ever came out of the big wardrobe upstairs. He was not much for parades and the like. He seldom mixed in politics, never became an enthusiastic church man. He neither smoked nor chewed, though he would take a beer every so often and once a year he liked to make wine down in the cellar. He did enjoy the Grand Army meetings, but that was largely because he enjoyed being with his friends. People liked him. He was a good storyteller, easygoing. He was also about as physically powerful as any man in town, and he was a very soft touch. He could lift a barrel full of sugar, which was considered quite a feat; but turning away a friend whose credit might not be the best seemed more than he was up to.

Twice in his life he had let friends have money when they came to him; twice he had suffered heavily from the loss. Fortunately for the Heisers, Mathilde was a determined and sensible wife. She was the manager of the two, and after George came back from Oil City she took charge. She kept the books, saw that he did not let much go on credit. Nor would she allow him to set any chairs out on the bare wooden floor. Otherwise, she said, his cronies would be sitting about half the day giving the place the wrong sort of appearance. They had at last been able to add a new window to the store front and appearances would be maintained so long as Mathilde had her say.

Mathilde's appearance was straightforward and intelligent. She had a fine head of dark-brown hair, a high forehead, and a set to her mouth that suggested she knew where she was going and that chances were good she would get there. She had had a considerable amount of education for a woman of that time and continued to keep up with her reading. Education was the thing, a proper education and hard work. It had been her way of life since childhood in Germany, and she intended to pass it on to the pride of her life, her son.

Victor was her second child. The first, a girl, had died of diphtheria. Victor had been struck down with it too, but he had been stronger. He looked much like his father now, only a ganglier, rawboned version. He was a serious, pink-faced boy, with big feet and blond hair, taller already than most men and far better educated. At sixteen he knew several languages, was well along in advanced mathematics, and had read about as widely as any boy in Johnstown. At his mother's insistence, his life was a steady round of school, homework, and being tutored in one extra course or another. If George Heiser had had his troubles making his way, that was one thing; Victor, Mathilde Heiser was determined, would not just get on, he would excel. There would be no going off to the Cambria works or coal mines for this young man, and no clerking behind a store counter either.

Now and again Victor had his moments away from all that. His father would step in and see that he got some time off. George Heiser had a wonderful way with children. He was forever telling them stories and listening to theirs. He took a great interest in his son and his schemes, one of which was to build a raft and float down the Conemaugh to the Allegheny, then on to the Ohio and Mississippi. Victor had taken some night classes in mechanical drawing at the library and had worked up plans on how the raft should be built. His idea was to catch the Conemaugh when it was high, otherwise he knew he would run aground.

In summer he would bring home accounts of his long rides out of town to the open country above the valley. Once he and one of his friends had gone all the way to the South Fork dam to take a look at the lake and the summer colony, but they had been sent on their way by the grounds keeper and never got to see much.

Other nights Victor talked about walking to the edge of town to watch the big summer revival meeting. He loved the powerful singing and the whooping and hollering of people "getting religion." On the way home he and his friends would try to imitate what they had seen, laughing and pounding each other on the back as they came along the streets. This was the kind of education George Heiser understood. He had grown up in Johnstown himself. His people had been among the Pennsylvania Germans who first settled the valley.

After ten the Heiser store was closed for the day, the lights out downstairs. When the downpour began, George and Mathilde did not think much of it, except that there would almost certainly be high water in the morning. But the thought bothered them very little, except for the inconveniences there might be. They listened to the rain drum on the roof and were glad to be inside.

If there was such a thing as a typical married couple in Johnstown on the night of May 30, 1889, George and Mathilde Heiser would come about as close as any to qualifying. Together, like Johnstown itself, they combined an Old World will to make good in the New with a sort of earlier-American, cracker-barrel willingness to take life pretty much as it came. Unlike a large number of Johnstown people, they were not directly beholden to the Cambria Iron Company, but their fortunes, like those of the entire valley, depended nonetheless on how red those skies glowed at night.

They had suffered the death of a child; they had tried their luck elsewhere and had lost. They fought dirt daily, saved every spare nickel, and took tremendous pride in the progress they were making. All things considered, Johnstown seemed a good place to be. It was their home.

II

Sailboats on the mountain

-1-

The lake had several different names. On old state maps it was the Western Reservoir, the name it had been given more than forty years earlier when the dam was first built. It was also known as the Old Reservoir and Three Mile Dam, which was the most descriptive name of the lot, if somewhat misleading, since the lake was closer to two than three miles long. The Pittsburgh people who had owned it now for ten years, and who had made a number of changes, called it Lake Conemaugh. But in Johnstown, and in the little coal towns and railroad stops along the way to Johnstown, it was generally known as South Fork dam.

South Fork was the nearest place to it of any size. Something like 1,500 people lived there in gaunt little frame houses perched on a hill just back from the tracks and the place where South Fork Creek flows into the Little Conemaugh River. Green hills closed in on every side; the air smelled of coal dust and pine trees. It was a town like any one of a half dozen along the main line of the Pennsylvania between Altoona and Johnstown; except for July and August, when things picked up considerably in South Fork.

The Pittsburgh people were coming and going then, and they were something to see with their troops of beautiful children, their parasols, and servants. Two or three spring wagons and buggies were usually waiting at the depot to take them to the lake. On Saturdays and Sundays the drivers were going back and forth several times a day.

The ride to the lake was two miles along a dusty country road that ran through the woods beside South Fork Creek, past Lamb's Bridge, then on up the valley almost to the base of the dam.

Seen from down below, the dam looked like a tremendous mound of overgrown rubble, the work of a glacier perhaps. It reared up 72 feet above the valley floor and was more than 900 feet long. Its face was very steep and covered with loose rocks. There were deep crevices between the rocks where, as late as May, you could still find winter ice hiding; but wild grass, bushes, and saplings had long since taken root across nearly all of the face and pushed up vigorously from between the rocks, adding to the overall impression that the whole huge affair somehow actually belonged to the natural landscape. There was hardly any indication that the thing was the work of man and no suggestion at all of what lay on the other side, except over at the far left, at the eastern end of the dam, where a spillway had been cut through the solid rock of the hillside and a wide sheet of water came crashing down over dark boulders. It was a most picturesque spot, and a favorite for picnics. Long shafts of sunlight slanted through a leafy gloom where the mountain laurel grew higher than a man could reach. And at the base of the falls a wooden bridge crossed the loud water and sent the road climbing straight to a clump of trees at the top of the dam, just to the right of the spillway.

There the road divided, with the left-hand fork crossing another long wooden bridge which went directly over the spillway. But carriages heading for the club took the road to the right, which turned sharply out of the trees into the sunshine and ran straight across the breast of the dam where, about a hundred yards out, the drivers customarily stopped long enough for everyone to take in the view.

To the right the dam dropped off a great deal more abruptly even than it had looked from below, and South Fork Creek could be seen glittering through the trees as it wound toward Lamb's Bridge.

On the other side of the road the bank sloped sharply to the water's edge, which was usually no more than six or seven feet below the top of the dam. From there the broad surface of the lake, gleaming in the sunlight, swept off down the valley until it disappeared behind a wooded ridge in the distance.

Along the eastern shore, to the left, were the hayfields and orchards of the Unger farm, neatly framed with split-rail fences. Beyond that was what was known as Sheep's Head Point, a grassy knoll that jutted out into the lake. Then there were one, two, three ridges, and the water turned in behind them, out of sight, running, so it seemed, clear to the hazy blue horizon off to the south.

At the western end of the dam the road swung on through the woods, never far from the water, for another mile or so, to the main grounds of the South Fork Fishing and Hunting Club, which, seen from the dam, looked like a colorful string of doll houses against the distant shore line.

From the dam to the club, across the water, was about a mile. Except for a few small coves, the narrowest part of the lake was at the dam, but there was one spot, on down past the club, where an east-west line across the water was nearly a mile. A hike the whole way around the shore was five miles.

When the water was up in the spring, the lake covered about 450 acres and was close to seventy feet deep in places. The claim, in 1889, was that it was the largest man-made lake in the country, which it was not. But even so, as one man in Johnstown often told his children, it was "a mighty body of water to be up there on the mountain."

The difference in elevation between the top of the dam and the city of Johnstown at the stone bridge was about 450 feet, and the distance from the dam to that point, by way of the river valley, was just under fifteen miles. Estimates are that the water of Lake Conemaugh weighed about 20 million tons.

The water came from half a dozen streams and little creeks that rushed down from Blue Knob and Allegheny Mountain, draining some sixty square miles. There was Rorabaugh Creek, Toppers Run, Yellow Run, Bottle Run, Muddy Run, South Fork Creek, and one or two others which seem never to have been named officially. South Fork Creek and Muddy Run were the biggest of them, but South Fork Creek was at least twice the size of the others. Even in midsummer it was a good twenty feet across. Like the others it was shallow, ice-cold, very swift, and just about a perfect place for trout fishing.

In South Fork there were scores of people who had been out on the dam and had seen the view. There were others who knew even more about the club and the goings on there because they worked on the grounds, tending lawns or waiting on tables at the clubhouse. But for everyone else the place was largely a mystery. It was all private property, and as the club managers had made quite clear on more than one occasion, uninvited guests were definitely not welcome.

The club had been organized in Pittsburgh in 1879. It owned the dam, the lake, and about 160 acres besides. By 1889 sixteen cottages had been built along the lake, as well as boathouses and stables. The cottages were set out in an orderly line among the trees, not very far apart, and only a short way back from the water. They looked far too substantial really to be called "cottages." Nearly every one of them was three stories tall, with high ceilings, long windows, a deep porch downstairs, and, often as not, another little porch or two upstairs tucked under sharp-peaked roofs. The Lippincott house with its two sweeping front porches, one set on top of the other, and its fancy jigsaw trim, looked like a Mississippi riverboat. The Moorhead house was Queen Anne style, which was "all the rage" then; it had seventeen rooms and a round tower at one end with tinted glass windows. And the Philander Knox house, next door, was not much smaller.

But even the largest of them was dwarfed by the clubhouse. It had enough windows and more than enough porch for ten houses. There were forty-seven rooms inside. During the season most of the club members and their guests stayed there, and the rule was that everyone had to take his meals there in the main dining room. where 150 could sit down at one time.

In the "front rooms" there were huge brick fireplaces for chilly summer nights, billiard tables, and heavy furniture against the walls. In summer, after the midday dinner, the long front porch was crowded with cigar-smoking industrialists taking the air off the water. String hammocks swung under the trees. Young women in long white dresses, their faces shaded under big summer hats, strolled the boardwalks in twos and threes, or on the arms of very proper-looking young men in dark suits and derbies. Cottages were noisy with big families, and on moonlight nights there were boating parties on the lake and the sound of singing and banjos across the black water.

In all the talk there would be about the lake in the years after it had vanished, the boats, perhaps more than anything else, would keep coming up over and over again. Boats of any kind were a rare sight in the mountains. There were rowboats on the old Suppes ice pond at the edge of Johnstown, and a few men had canoes along the river below the city. But that was about it. Not since the time when Johnstown had been the start of the canal route west had there been boats in any number, and then they had been only ungainly canal barges.

The club fleet included fifty rowboats and canoes, sailboats, and two little steam yachts that went puttering about flying bright pennants and trailing feathers of smoke from their tall funnels. There was even an electric catamaran, a weird-looking craft with a searchlight mounted up front, which had been built by a young member, Louis Clarke, who liked to put on a blue sailor's outfit for his cruises around the lake.

But it was the sailboats that made the greatest impression. Sailboats on the mountain! It seemed almost impossible in a country where water was always a tree-crowded creek or stream, wild and dangerous in the spring, not much better than ankle-deep in the hottest months. Yet there they were: white sails moving against the dark forest across a great green mirror of a lake so big that you could see miles and miles of sky in it.

Some of the people in Johnstown who were, as they said, "privileged" to visit the club on August Sundays brought home vivid descriptions of young people gliding over the water under full sail. It was a picture of a life so removed from Johnstown that it seemed almost like a fantasy, ever so much farther away than fifteen miles, and wholly untouchable. It was a picture that would live on for a long time after.

That the Pittsburgh people also took enormous pleasure in the sight seems certain. There was no body of water such as this anywhere near Pittsburgh. There were, of course, the Monongahela and Allegheny rivers, but they were not exactly clean any longer, and with the mills going full blast, which they had been for some time now, the air around them was getting a little more unpleasant each year. It was a curious paradox; the more the city prospered, the more uncomfortable it became living there. Progress could be downright repressive. But fortunately for the Pittsburgh people, it was very much within their power to create and maintain a place so blessed with all of nature's virtues.

This water was pure and teeming with fish, and the air tasted like wine after Pittsburgh. The woods were full of songbirds and deer that came down to drink from the mist-hung lake at dawn. There were wild strawberries everywhere, and even on the very hottest days it was comfortable under the big trees.

In fact, with its bracing air, its lovely lake, and the intense quiet of its cool nights, the South Fork Fishing and Hunting Club must have seemed like paradise after Pittsburgh. Under such a spell even a Presbyterian steel master might wish to unbend a little.

The summer resort idea was something new for that part of the country. And only the favored few had the time or the money to experiment with it. But so far every indication was that the club was a great success. There had been problems from time to time. Poachers had been a continuing nuisance. The summer of 1888 had been cut short when a scarlet-fever scare sent everyone packing off home to Pittsburgh. Still, everything considered, in 1889 it looked as though the men who had bought the old dam ten years earlier knew what they were doing.

-2-

The first member of the South Fork Fishing and Hunting Club to take an interest in the regenerative powers of the Alleghenies was Andrew Carnegie.

Carnegie had been going to what he called "The Glorious Mountain" long before the South Fork organization was put together. He had his own modest frame house at Cresson, which was one of the first summer resorts in Pennsylvania and the only one of any consequence in the western part of the state. It was owned by the Pennsylvania Railroad and was located fourteen miles up the line from South Fork at the crest of the Allegheny range.

Cresson, or Cresson Springs as it was also known, had been started before the Civil War by a doctor named Robert Montgomery Smith Jackson. The main attractions at Cresson, aside from the mountain air and scenery, were the "iron springs," the best-known of which was the Ignatius Spring, named after "the venerable huntsman" Ignatius Adams, who first discovered its life-preserving powers and whose ghost was said still to haunt the place. According to Jackson, "by drinking this water, dwelling in the woods and eating venison," Ignatius had "lived near the good old age of one hundred years." Jackson was against whiskey, slavery, and what he called the "present tendency to agglomerate in swarms, or accumulate in masses and mobs." Those "gregarious instincts [which] now impel this race to fix its hopes of earthly happiness on city life alone" would, he was convinced, be the undoing of the race. Life in the country was the answer to practically every one of man's ills, and particularly life on the Allegheny Mountain.

Jackson's ambition ("a mission, solemn as a command from Heaven," he called it) was to make Cresson "the place of restoration for all forms of human suffering." He got his friend J. Edgar Thomson, president of the Pennsylvania, interested, and the railroad built a hotel and developed the place, though, as things turned out, along rather different lines. Carnegie, B. F. Jones, and a few other Pittsburgh businessmen, none of whom seems to have been suffering very much, built cottages and the summer trade flourished. Every passenger train bound east or west stopped there. Well-to-do families from Pittsburgh and Philadelphia arrived for summer stays of several weeks.

Jackson meanwhile had set down the fundamentals of his philosophy along with a detailed natural history of the Allegheny highlands in a book called *The Mountain*. He borrowed heavily from Wordsworth and Thoreau, and, in his own way, did about as much as anyone to sum up the wild beauty of the area. Also, in his spare time, he tended bar at the hotel and would be remembered for years after for the two jars he kept prominently displayed on one shelf, flanked on either side by whiskey bottles. In each jar, preserved in alcohol, was a human stomach. One had belonged to a man who had died a natural death, and was, according to all who saw it, an exceedingly unappetizing sight. But it was, nonetheless, an improvement over its companion piece, which, according to its label, had belonged to a man who had died of delirium tremens. When setting out drinks, the doctor seldom failed to call attention to his display. The result was that his bar became the best patron-ized of any for miles about. Regular customers grew quite attached to the jars; word of them spread far, and along with the iron springs, they appear to have been a major attraction at Cresson for several years.

Jackson served in the war; then, in 1865, despite all his good life on the mountain, he died at the age of fifty.

By 1881, to accommodate the growing trade, the railroad cleared space in a maple grove and built another hotel, The Moun-tain House, which, with its endless thick-carpeted halls, its many towers and flowing stairways, was easily the grandest piece of ar-chitecture in Cambria County.

Carnegie's house, only a short distance from the hotel, was his only real home in Pennsylvania by that time. Though no one held more sway in Pittsburgh, he had not lived there for nearly twenty years. He visited often, and was front-page news when he did, but the rest of the time he was either in New York, Scotland, or at Cresson. He loved Cresson and talked up its charms with great vigor, which is perhaps not surprising for a Scot who had been brought up on Burns and had learned to quote him before he had learned to read. ("My heart's in the Highlands, my heart is not here;/ My heart's in the Highlands a-chasing the deer.") He courted his young wife there in the summer of 1886. ("A.C. walked home with me in the starlight . . ." she wrote in her diary at Cresson. "Such wonderful happiness . . . the happiest day of my life.") He also very nearly died there of typhoid later that fall; and his mother, who took sick at the same time and lay in the next room down the hall, did die there on November 10. He entertained his distinguished friends at Cresson (Matthew Arnold had stopped over in 1883), and he managed, as was his pleasure, to keep his mind free and above the petty preoccupations of the steel business.

As one admiring writer of the period explained it: "All other iron and steel magnates, with the exception of Carnegie, lived in Pittsburgh and were swayed constantly by the local gossip, by the labour troubles, and by the rumours of competition and low prices that floated from office to office. To-day they were elated; to-morrow they were depressed. To-day they bought; to-morrow they sold. Carnegie, on the other hand, deliberately placed himself where these little ups and downs were unnoticed. . . . the news that Coleman had quarrelled with Shinn, or that coke-drawers wanted five cents a day more, was of small consequence. One thing he knew—that civilisation needed steel and was able to pay for it. All else was not worth troubling about."

The place had its hold on him. He went on bird walks; he read; he talked and talked and talked. And if the views and good company were not enough, there were the "curative powers," as the guidebook described them, of the iron springs. And who might more readily endorse such a tonic than the bouncy little ironmaster himself?

But Cresson had its drawbacks. It was a public place for one, sitting almost on top of the railroad. For another, there was no water. The story goes that there was (and still is) a house at Cresson where the rain water off of one roof eventually ends up in the Atlantic Ocean, while the rain water off the other finds its way to the Mississippi Valley and the Gulf of Mexico. The resort was at the very crest of the Allegheny divide, and though springs were plentiful, every bit of water there was drained off in both directions.

As a result, except for drives and walks, there was really not much to do at Cresson. Tennis had not yet caught on there. Golf, which was to one day be the great passion of Carnegie and his kind, would not even be attempted in the United States until 1888. But boating was already distinctly fashionable, and fishing had been coming more and more into its own as a gentleman's sport.

Part of the increasing appeal of fishing seemed to be the multitude of trappings it called for. Where once the well-equipped angler needed only the simplest and most inexpensive sort of gear, now in the late 1880's a whole line of elaborate and expensive paraphernalia was said to be necessary. Bait boxes, boots, collapsible nets, cookstoves, silk line, creels, reels and casting rods that cost as much as twenty dollars, even costly books on the subject, were the sign of the true sportsman. All of which seemed to make fishing, and particularly trout and bass fishing, which were generally referred to as a "science," that much more attractive to the man of means. His interest in the sport not only showed his love of the great outdoors, but also that he had both the money and the brains to participate.

Of course, too, few pastimes there were which would take a man so far, in spirit at least, from the rude industrial grind. "God never did make a more calm, quiet, innocent recreation than angling," were the words of Izaak Walton in the newly reissued, and quite costly, two-volume edition of his great work on the subject. One Walton disciple of the 1880's, who as it happens was a Vermonter, wrote about this time, "I take my rod this fair June morning and go forth to be alone with nature. No business cares, no roar of the city, no recitals of others troubles . . . no doubts, no fears to disturb me as, drinking in the clear, sweet air with blissful anticipation I saunter through the wood path toward the mountain lake." Had he been a Pittsburgh steel man he need only have added "no competitors, no labor agitators."

So all that was needed to improve on Cresson was enough water for boating and fishing. A mountain lake, in short; plus some privacy. It would also be well to be back a way from the railroad, though not too far back, as the railroad was the one and only way to make the trip from Pittsburgh. As it was, the ride took about an hour and fifteen minutes to Johnstown, then another twenty minutes to come on up the mountain. That was about long enough. It would be best, therefore, to be somewhere in the same general area.

The old reservoir above South Fork certainly must have seemed a perfect solution. It answered every need: It was well back from the railroad; it was only a matter of miles from Cresson; and South Fork Creek was well known as one of the best trout streams in the state. True, the dam needed a great deal of repair work after so many years of neglect, but that could be handled all right, and especially since the property could be had for such a good price. Or so must have run the reasoning of the Pittsburgh men who bought it in 1879. Carnegie was not among them at this particular stage, but no doubt his presence so nearby on the mountain added still another enticing attraction to the scheme.

The prime promoter behind the move was a onetime railroad tunnel contractor and now-and-then coke salesman and real-estate broker by the name of Benjamin F. Ruff.

Ruff bought the dam and the lake from John Reilly, an Altoona Democratic politician and former Pennsylvania Railroad official who was then serving what would be his only term in Congress. Ruff paid $2,000 for the property, which was $500 less than Congressman Reilly had paid for it four years earlier when he had bought it from the Pennsylvania.

Ruff then rounded up fifteen other Pittsburgh gentlemen who each, with one exception, bought a single share in the operation for $200. The exception was one of the most interesting young men in Pittsburgh.

In the spring of 1879, when all this was going on, Henry Clay Frick was only twenty-nine years old, solemn, enigmatic, strikingly handsome, and already worth an even million. The grandson of wealthy, old Abraham Overholt, the Mennonite whiskey maker, Frick had made his own fortune in the coke business, and largely through his dealings with Pittsburgh's number-one coke consumer, Andrew Carnegie.

Frick bought three shares. Ruff kept four for himself. A charter was drawn up stating that the name of the organization was to be the South Fork Fishing and Hunting Club of Pittsburgh, that its "object" was to be "the protection and propagation of game and game fish, and the enforcement of all laws of this state against the unlawful killing or wounding of the same." It was also stated that the club's place of business was to be Pittsburgh, in Allegheny County, not in Cambria County where the property was located.

On November 15, 1879, the charter was approved and signed in the Court of Common Pleas in Allegheny County by Judge Edwin H. Stowe, who for some unknown reason ignored the provision in the law which called for the registration of a charter in the "office for recording in and for the county where the chief operations are to be carried on." Nor did the sportsmen make any effort to conform to the law. Perhaps it seemed a minor point and was overlooked by mistake. In any case, the charter was secured without the knowledge of the authorities in Cambria County, and there would be speculation for years to come as to what might have happened right then and there had they and Judge Stowe gone about their business in strict accordance with the rules.

Ruff was to be the president of the new club. The capital stock was to be $10,000, but that was soon increased to $35,000 when it became known how much work was needed to get the dam in shape. For in 1879 the South Fork dam was getting on in years, and the years had been hard on it.

Forty-three years earlier, in 1836, the legislature of the state of Pennsylvania had approved funds for the building of a reservoir on the western slope of Allegheny Mountain to supply extra water during dry months for the new canal system from Johnstown to Pittsburgh. The first appropriation was for $30,000, but before the project was finished nearly $240,000 would be put into the dam; and two years after it was finished the whole thing would be obsolete and of no use whatsoever.

The canal from Johnstown to Pittsburgh was known as the Western Division of the state's "Main Line" canal, which had been built to compete with New York's thriving, new Erie Canal. The Western Division officially opened for business in May 1831, when a Johnstown barge pulled into Pittsburgh after traveling 104 miles in less than forty-eight hours.

The Eastern Division, running from Philadelphia to Hollidaysburg, at the foot of the eastern slope of Allegheny Mountain, was opened the next year. The only thing that then remained to be finished was an ingenious system of railroads and steam-powered hoists designed to get the boats up and over the mountain. But by spring of 1834 that too had been built and was open to traffic. For the first time the wharves of Philadelphia had a direct, nonstop link with the headwaters of the Ohio. It had cost a staggering sum; the state was nearly bankrupt; but that old, formidable barrier to the Pennsylvania route west, Allegheny Mountain, had been bested, and the general course of the country's epic push to the Mississippi and beyond had been set for years to come.

The system devised for crossing the mountain was widely hailed as one of the engineering wonders of the age. Within a distance of thirty-six miles it overcame an increase in elevation of nearly 1,400 feet, or about twice the elevation the Erie Canal had to overcome along its entire length of 352 miles. Known as the Portage Railroad, it included a series of five inclined planes on each side of the mountain, ten in all, connected with a narrow-gauge railroad. Barges, passengers, freight, everything was hauled up one side and let down the other with hemp ropes thick as a man's leg. It was a thrilling experience for travelers, a goodly number of whom chose to go by way of the Pennsylvania, rather than the Erie Canal, for that very reason.

Charles Dickens, one such traveler, described the ride in his *American Notes:*

> It was very pretty travelling thus, at a rapid pace along the heights of the mountain in a keen wind, to look down into a valley full of light and softness: catching glimpses, through the tree-tops, of scattered cabins; children running to the doors; dogs bursting out to bark, whom we could see without hearing; terrified pigs scampering homewards; families sitting out in their rude gardens; cows gazing upward with a stupid indifference; men in their shirt-sleeves looking on at their un-finished houses, planning out tomorrow's work; and we riding onward, high above them, like a whirlwind.

The adventurous journey also included, just to the east of Johnstown, a ride through the first railroad tunnel in the country and, in Pittsburgh, a ride across the first suspension bridge, an aque-duct over the Allegheny River designed by German-born John Augustus Roebling, who would later conjure up that wonder of wonders of the 1880's, the Brooklyn Bridge.

But from Johnstown west the canal was troubled by water shortages nearly every summer. Operations were interrupted. Busi-ness suffered at a time when business had to be especially good to make up for winter, when virtually every moving thing stopped in the mountains for weeks on end, and spring, when the floods came. Particularly troublesome was the canal basin in Johnstown. Despite all the water that rushed into the valley in springtime, along toward mid-July the basin came close to running dry.

The solution seemed obvious enough. Put a dam in the moun-tains where it could hold a sufficient supply of water to keep the basin working and the canals open, even during those summers when creeks vanished and only weeds grew.

Work began on the Western Reservoir above South Fork in 1838, after some 400 acres had been cleared of timber. The site had been selected and surveyed by Sylvester Welsh, head engineer for the canal. He proposed an earth dam of 850 feet in length, with a spillway at one or both ends "of sufficient size to discharge the waste water during freshets, and sluices to regulate the supply for the canal." It was also important, he said, that the bed of the spill-way be solid rock and that no water be permitted to pass over the top of the dam. The design of the dam was worked out by a young state engineer named William E. Morris, who approved the location because, as he stated in a report made in 1839, it was in an area where there was enough drainage to provide a "certain" supply of water. He too proposed an earth dam 850 feet across the top and 62 feet high. He estimated that it would take a year to do the job.

The contractors chosen were James N. Moorhead of Pittsburgh and Hezekiah Packer of Williamsport. According to lengthy studies made by civil engineering experts years later, they did a competent job. Certainly they went about it with considerable care and patience and despite continuing delays. For, as it turned out, fifteen years passed before the dam was finished.

In 1842 work was halted because the state's finances were in such bad shape that there was simply no more money to continue the job. For the next four years nothing was done. Then when the work did start again, it was only for another two years. A local cholera epidemic caused "a general derangement in the business," until 1850, when the project again resumed, and for the final time.

The construction technique was the accepted one for earth dams, and, it should be said, earth dams have been accepted for thousands of years as a perfectly fine way to hold back water. They were in fact the most common kind of dam at the time the South Fork work began and they were the most economical. The basic construction material was readily available at almost any site, it was cheap, and it required a minimum of skilled labor. Virtually any gang of day laborers, and particularly any who had had some experience working on railroad embankments, was suitable. But since the basic raw material, earth, is also highly subject to erosion and scour, it is absolutely essential that a dam built of earth, no matter how thick, be engineered so that the water never goes over the top and so that no internal seepage develops. Otherwise, if properly built and maintained, an earth dam can safely contain tremendous bodies of water.

The South Fork embankment was built of successive horizontal layers of clay. They were laid up one on top of the other after each layer had been packed down, or "puddled," by allowing it to sit under a skim of water for a period of time, so as to be watertight. It was a slow process. And as the earth wall grew increasingly higher, it was coated, or riprapped, on its outer face with loose rocks, some so huge that it took three teams of horses to move them in place. On the inner face, which had a gentler slope, the same thing was done, only with smaller stones.

The spillway, as Welsh had stipulated, was not cut through the dam itself, but through the rock of the hillside to which the eastern end of the dam was "anchored." The spillway was about 72 feet wide. The over-all length of the breast was just over 930 feet. The width on top was about 20 feet. The thickness at the base was some 270 feet.

At about the exact center of the base, there were five cast-iron pipes, each two feet in diameter, set in a stone culvert. They were to release the water down to South Fork, where it would flow on to the Johnstown basin by way of the Little Conemaugh. The pipes were controlled from a wooden tower nearby. On June 10, 1852, the work on the dam was at last completed; the sluice pipes were closed and the lake began to fill in. By the end of August the water was 40 feet deep.

But about the time the dam was being finished, J. Edgar Thomson, who was then chief engineer for the up-and-coming Pennsylvania Railroad, was making rapid progress with his daring rail route over the mountains, which included what was to become famous as the Horseshoe Curve. The canal was about to be put out of business.

The Pennsylvania was racing to complete a route west to compete with the New York Central, the Erie, and the B & O, which were each pushing in the same direction. The last part of the run, from Johnstown to Pittsburgh, was ready in late 1852. On December 10, six months after the South Fork dam had been finished, a steam engine made an all-rail run from Philadelphia to Pittsburgh. J. Edgar Thomson became president of the road about the same time, and the company was on its way to becoming within a very few years the biggest and far and above the most powerful single force in the state (and in the Statehouse); the biggest customer for nearly everything, but especially coal, iron, and steel; the biggest employer; and the biggest influence on the way people lived from one end of Pennsylvania to the other. By the end of the '80's it would be the mightiest of the nation's many mighty railroads.

The effect of the new railroad on the state's troublesome, costly, and beloved canal system was disastrous—almost immediately. Within two years after the railroad opened, the legislature voted to put the "Main Line" up for sale for not less than $10 million. Understandably there were no takers. The one likely prospect was the Pennsylvania itself, which could readily use the right of ways. Three years later the sale was made, with the Pennsylvania paying $7.5 million for the system, which included the Main Line, the Portage Railroad, and, as it happened, the South Fork dam.

Having no use for the dam, the railroad simply let it sit. Nothing whatsoever was done to maintain it. In fact, from 1857, the year the railroad took possession, until 1879, twenty-two years later when the Pittsburgh men took over, the dam remained more or less quietly unattended, moldering away in the woods, visited only once in a while by fishermen or an occasional deer hunter.

And it was only five years after the state sold it to the Pennsylvania that the dam broke for the first time.

In the late spring of 1862, about the time the Union Army under McClellan was sweating its way up the blazing Virginia peninsula, for a first big and unsuccessful drive on Richmond, the mountains of Pennsylvania were hit by heavy thunderstorms. Hundreds of tiny creeks and runs and small rivers went roaring over their banks, and in Johnstown the *Tribune* ran the first of its musings on what might be the consequences should, by chance, the dam at South Fork happen to let go. Eight days later, on June 10, the dam broke.

The break was caused by a defect in the foundation near the stone culvert. The accepted theory locally was that various residents had been stealing lead from the pipe joints during the years the dam had been abandoned, that serious leaks had been the result, and that the break had come not long after. Exactly how big the break was is not known, as no records were made and no photographs were taken. The important fact was that though there was much alarm in the valley below the dam, the break caused little damage since the lake was less than half full, the creeks were low, and a watchman at the dam, just before the break, had released much of the pressure by opening the valves. (It was also somewhere along about this time that the wooden tower for controlling the discharge pipes caught fire and burned to the ground.)

From then on until the Pittsburgh sportsmen appeared on the scene seventeen years later the lake was no lake at all, but little more than an outsize pond, ten feet deep at its deepest point. At the southern end, grass quickly sprouted across acres and acres of dried-up lake bed and neighboring farmers began grazing their sheep and cattle there.

In 1875 Congressman Reilly, who had spent most of his working life with the Pennsylvania in nearby Altoona, and who must have thereby known about the dam for some time, bought the property and, like the Pennsylvania, did nothing with it. He just held on to it, apparently on the look for another buyer, which he found four years later in Benjamin Ruff. But before selling at a slight loss to Ruff, he removed the old cast-iron discharge pipes and sold them for scrap.

Ruff's idea of what to do about the dam was relatively simple and seemed realistic enough at first. He would rebuild it to a height of only forty feet or so and cut the spillway down some twenty feet deeper to handle the overflow. But when he found that this would cost considerably more than repairing the old break and restoring the dam to somewhere near its original height, he chose the latter course.

The first indication in Johnstown and thereabouts that a change was in the offing above South Fork was an item in the *Tribune* on October 14, 1879. "Rumors" were reported that a summer resort was to be built by a Western Game and Fish Association. The next day there was a notice calling for fifty men to work, but no name of the organization was given.

For some reason or other, intentionally or otherwise, the Pittsburgh men kept the correct name of their organization from receiving any kind of public notice. It was a course of action which would later be interpreted as evidence that they had had no desire for anyone to come looking into their business in general, or their charter in particular.

Ruff set about repairing the dam by boarding up the stone culvert and dumping in every manner of local rock, mud, brush, hemlock boughs, hay, just about everything at hand. Even horse manure was used in some quantity. The discharge pipes were not

replaced, and the "engineering" techniques employed made a profound impression on the local bystanders.

The man immediately in charge of this mammoth face-lifting
was one Edward Pearson, about whom little is known except that
he seems to have been an employee of a Pittsburgh freight-hauling
company that did business with the railroad and that he had no
engineering credentials at all.

The entire rebuilding of the dam ended up costing the club
about $17,000, and there was trouble from the start. On Christmas
Day, 1879, only a month or so after work had begun, a downpour
carried away most of the repairs. Work was discontinued until the
following summer. Then, less than a year later, in February of
1881, once again heavy rains caused serious damages.

No one seems to have been particularly discouraged by all this,
however. Along toward the end of March the lake was deep
enough for the clubmen to go ahead with their plan to stock it. The
first of the small steamboats was being assembled and the clubhouse
was close to being readied for the grand opening. In early June the
fish arrived by special tank car from Lake Erie, 1,000 black bass,
which ended up costing the club about a dollar apiece by the time
the last expenses were paid. According to the *Tribune*, which noted
these and all other bits of news it could uncover concerning the
club, only three of the fish died, "one of which was a huge old
chap, weighing over three pounds."

The *Tribune* had also reported earlier that the Pennsylvania
was planning to build a narrow-gauge spur from South Fork to the
lake and that the clubmen were shopping about for land downriver
from Johnstown where they intended to establish a private deer
park of 1,500 acres. Neither claim was true, but both seemed perfectly reasonable and fitted in with the picture most Johnstown
people had of the club and its members. Would not even the high
and mighty Pennsylvania Railroad gladly provide any number of
special conveniences for the likes of such men? Was not a deer park
a fitting aristocratic touch for their new mountain domain? Certainly money was no problem. Were not the members of the club
millionaires to the man?

The plain truth was that a goodly number of them were; quite
a few of them, however, were not, and two or three of them were a
great deal better than millionaires. But by the standards of most

men, they were, every last one of them, extraordinarily rich and influential. Yet, one of the curious things about the club is that the make-up of its membership, exactly who was who at the lake, was not generally known around Johnstown. If the club appears to have been rather cozy about its name, it was even more so about publicizing who belonged to it. Not until well after the events of May 31, 1889, was a full list of the membership published publicly. And quite a list it was.

The membership of the South Fork Fishing and Hunting Club, according to its initial plans, was never to exceed one hundred sportsmen and their families. The membership fee was $800. There was to be no shooting on Sundays; and those members who did not have cottages of their own were limited to a two-week stay at the clubhouse. As the summer season of 1889 was approaching, there was a total of sixty-one names on the membership roster.

The South Fork Fishing and Hunting Club, it should be kept in mind, was a most unostentatious affair by contrast to such watering spots of the time as Newport, Cape May, or the lavish new lakeside resort in New York, Tuxedo Park. There was no opulence. There were no liveried footmen, no Tirolean-hatted gamekeepers such as at Tuxedo, no "cottage" architecture to approach the likes of Newport. There was not even a comparison to be made, unless the South Fork group was to be measured by the per capita worth of its members—or the industrial and financial power they wielded —which, everything considered, was often the way such things were measured. On that basis the little resort on Lake Conemaugh was right in the same league.

One Pittsburgh newspaper called it the "Bosses Club," and aptly so. Carnegie's name by itself on the membership list would have been reason enough. And the same holds for Henry Clay Frick, for much had happened to the young "Coke King" since he had first joined with Benjamin Ruff to launch the club.

In 1881, while in New York, Frick had stopped by the Windsor Hotel on Fifth Avenue to pay a call on Carnegie and his mother and to talk a little business. (Frick happened to be on his honeymoon at the time, but he was not the man to let that, or a·ything else, stand in the way of progress.) When the meeting was over, he and Carnegie were partners in the coke trade, and from then on it did even better than before. By 1889 the H. C. Frick Coke Company was capitalized at $5 million; it owned or controlled 35,000 acres of coal land and employed some 11,000 men. Moreover, in January of that year, Frick had been made Chairman of Carnegie, Phipps & Company, which meant that he was commander in chief of the whole of the Carnegie iron and steel enterprises, which were by then the biggest in the world.

Carnegie by this time had said something to the effect that he was not much interested in making more money and was spending no more than six months a year at his business. He wanted someone who could manage things. In Frick, whom he did not especially like and whom no one seemed quite able to fathom, he found exactly the right man. By 1889 this humorless, solitary, complex son of a German farmer, who was then still six months from turning forty, was the most important man in Pittsburgh.

Along with Frick, the club roster included Henry Phipps, Jr., Carnegie's partner since the earliest days of the business, before the Civil War. A pale, painstaking man who stood no more than an inch taller than the five-foot-two Carnegie, Phipps was the financial wizard of Carnegie, Phipps & Company. In the early days he had won certain acclaim for his ability to borrow money for the struggling ironworks and for an old horse he owned, which, according to Pittsburgh legend, was capable of taking him on his rounds of the banks without any guidance whatsoever. By 1889, however, Phipps was one of the three or four top men in the steel business and one of the wealthy men of the country.

Besides the big three—Carnegie, Frick, and Phipps—the Carnegie empire was also represented at South Fork by John G. A. Leishman, the vice-chairman of the firm, and by Philander Chase Knox, a bright little sparrow of a man, who was the company's number-one lawyer as well as the personal counsel for both Carnegie and Frick.

Then there were Robert Pitcairn and Andrew Mellon, two excellent friends to have if you were doing business in Pittsburgh. Pitcairn ran the Pittsburgh Division of the Pennsylvania Railroad, which, as jobs went in those days, was a most lofty position indeed. In terms of pure power and prestige, few men outranked him.

Mellon was a shy, frail-looking young man, still in his thirties, and exceedingly quick-witted. Along with his father, old Judge Mellon, he was in the banking business, T. Mellon & Sons, and up to his elbows in the financial doings behind much of Pittsburgh's furious industrial growth. It had been the Mellons who lent Henry Frick the money to buy his first coal land, and who backed him again (with something like $100,000) during the panic of 1873 when he wanted to buy up still more. (One story has it that Judge Mellon, who was a staunch Methodist, made the loan only after Frick implied that he would use his influence to have the Overholt distillery shut down.) Frick and young Andrew were not far apart in age and became fast friends, traveling to Europe together, dealing in business for years, but never calling each other anything but Mister Frick and Mister Mellon.

James Chambers and H. Sellers McKee ran what they claimed was the largest window-glass works on earth. Durbin Horne and C. B. Shea ran Pittsburgh's leading department store, Joseph Horne and Company. D. W. C. Bidwell sold DuPont blasting powder for coal mining. Calvin Wells, A. French, James Lippincott, and John W. Chalfant were in the steel business in one way or other.

And so the list went. It included names from the Pittsburgh "Blue Book" (Thaw, Laughlin, McClintock, Scaife) and from the lists of directors of several Pittsburgh banks (Schoonmaker, Moorhead, Caldwell). There were among them the founders of the city's new business club, the Duquesne Club, and the new preparatory school for young men, Shady Side Academy.

There were also among them, it is interesting to note, a future Secretary of the Treasury (Mellon, who would serve under Harding, Coolidge, and Hoover) and a future Secretary of State (Knox, who would be Attorney General under McKinley and Theodore Roosevelt before taking over the State Department under Taft). There was one future Congressman (George F. Huff, who was a banker and coal operator), a future diplomat (John G. A. Leishman, who would be America's first Ambassador to Turkey), and a future President of the Pennsylvania Railroad (Samuel Rea).

Frick and Mellon would not only go on to amass fortunes of spectacular proportions, but would also demonstrate surprisingly good taste in putting together two of the world's finest private art collections. Carnegie, who was already worth many millions, would wind up with more money than any other American except old John D. Rockefeller and would give away well over $300 million of it with no little fanfare.

In 1889, however, all that was still a good way off. The Carnegie whiskers had not as yet turned their glistening white. No palaces had been built for Carnegie or Frick on New York's Fifth Avenue. No Rembrandt's had been bought, no daughters married off to titled Europeans, no parks donated, no foundations established. Carnegie thus far had built only one of his free libraries, in Braddock, near Pittsburgh and the site of his gigantic Edgar Thomson works, which he had named after his old friend Thomson, who also happened to be his best customer. At the dedication of the Braddock library in March of 1889, Carnegie used the opportunity to say that he would certainly like to build another one in Homestead someday, but that there had been really far too much labor trouble over there. The implication was pretty obvious and suggested that the purpose behind these earliest benefactions at least may not have been altogether altruistic.

Good works, public service, and any ideas about giving away surplus money were all still things of the future.

It was in fact during the month of May 1889 that Carnegie was finishing up a magazine article to become known as "The Gospel of Wealth," in which he said, and much to the consternation of his Pittsburgh associates, "The man who dies thus rich dies disgraced." The gist of the article was that the rich, like the poor, would always be with us. The present system had its inequities, certainly, and many of them were disgraceful. But the system was a good deal better than any other so far. The thing for the rich man to do was to divide his life into two parts. The first part should be for acquisition, the second for distribution. At this stage the gentlemen of the South Fork Fishing and Hunting Club were attending strictly to the first part. Business was the overriding preoccupation for now, and business in Pittsburgh, either directly or indirectly, meant the steel business, which in 1889 was doing just fine.

True, orders for steel rails had tapered off some. Breakneck railroad building in the west had meant palmy days in the accounting offices along the Allegheny and Monongahela. Seventy-three thousand miles of track had been put down in the 1880's, or more than twice as much track as there had been in the whole country when the war ended at Appomattox. Steel production had more than tripled during those ten years. But now there were also orders for all sorts of architectural steel, huge beams and girders for totally new kinds of buildings called "skyscrapers" which were going up in Chicago. In 1887 the Government had made its first order for American steel for ship armor. A new kind of Navy was being built, a steel Navy, including, in 1889, a battleship called the *Maine*, the biggest thing ever built by the Navy, for which Carnegie, Phipps & Company was making the steel plates.

And if the rail business was not quite as good as it had been, the United States was, nonetheless, producing about two tons of rails for every one made in England. As a matter of fact, for three years now, the United States had been the leading steel producer in the world. Pittsburgh, which before the war had not had a single mill as big as those in Johnstown, was now throbbing like no other industrial center in the land. The sprawling complex of mills in Chicago had been providing serious competition of late. Prices were not as high as the steel men would have liked to see them (they never were), and there was more talk among them every week about cutting wages. But if the labor leaders could be dealt with (and there was no reason to think they could not be), then there was every reason to believe that Pittsburgh would keep booming for years.

As far as the gentlemen of the South Fork Fishing and Hunting Club were concerned no better life could be asked for. They were an early-rising, healthy, hard-working, no-nonsense lot, Scotch-Irish most of them, Freemasons, tough, canny, and, without question, extremely fortunate to have been in Pittsburgh at that particular moment in history.

They were men who put on few airs. They believed in the sanctity of private property and the protective tariff. They voted the straight Republican ticket and had only recently, in the fall of 1888, contributed heavily to reinstate a Republican, the aloof little Harrison, in the White House. They trooped off with their large families regularly Sunday mornings to one of the more fashionable of Pittsburgh's many Presbyterian churches. They saw themselves as God-fearing, steady, solid people, and, for all their new fortunes, most of them were.

Quite a few had come from backgrounds as humble as Carnegie's. Phipps and Pitcairn were Scotch immigrants who had been boyhood pals with Carnegie in what was known as Slabtown, one of the roughest sections of Allegheny, across the river from Pittsburgh. Leishman grew up in an orphanage. Frick, despite the wealth of grandfather Overholt, had started out in business with little more than a burning desire to get rich.

They and the others were now living in cavernous, marble-floored houses in the new East End section of Pittsburgh. Several made regular trips to Europe, and those who did not always stopped at the finest hotels in New York or wherever else they went. They now considered themselves, each and all, as among the "best people" in Pittsburgh. They pretty well ran the city. They were living the good life as they thought the good life ought to be lived. But never for very long did they take their eye off the real business of the human condition as they saw it—which was business. That they should spend some time together in the summer months, away from Pittsburgh, but not too far away, mind you, seemed, no doubt, a perfectly natural extension of the whole process.

The reaction in Johnstown to their doings on the mountain was mixed. That the Pittsburghers with all their money should think enough of the country around Johnstown to want to summer there was, of course, terribly flattering. As far back as the 1850's there had been some discussion in Johnstown about the area's potential as a summer resort. One *Tribune* editor decided to set forth Johnstown's charms in no uncertain terms. "Our scenery is grand beyond description," he wrote, "the atmosphere cool, and invigorating; trout in the neighboring streams large and numerous; drives good; women beautiful and accomplished; men all gentlemen and scholars, hotels as good as the best." For all anyone knew, the South Fork venture might lead to other resort developments in the area. The valley might become famous; property values would mount. It was not an unpleasant thought.

The club had already provided a lot of work in the South Fork area, and much to the irritation of the club's management it had also provided some excellent sport for local anglers. Slipping onto the property in the early morning or toward sundown was no problem for a local man or boy. The well-stocked lake and streams provided any number of suppers in the neighborhood from the time the dam was first restored. The grounds were well posted, but that discouraged nobody especially. If anything, it only added to the fun.

When the club responded by putting in fences over the best

trout streams, the fences mysteriously disappeared. Relations then deteriorated fast, with the club authorities threatening to shoot any invaders caught on the grounds after dark. During a flood in 1885, one farmer named Leahy, whose property adjoined the lake, decided to rent out fishing space on some of his submerged acreage. The clubmen said he ought not to do that and threatened to take him to court. This had no effect on Leahy, so they tried to buy him out, but he said he was not interested in selling. It was only after an intermediary was brought in, and lengthy negotiations transacted, that the farm was purchased for $4,000 and the fish thereby further protected.

Any love lost locally by such tactics apparently bothered the club management not in the least. A classic undeclared war between poacher and country squire went on for years. It never became a shooting war, despite the threats, but it did leave widespread resentment in the area around the lake that would one day come back to haunt the clubmen.

Far more important, however, was the way people felt about the dam.

Even before the first full season at South Fork got under way in 1881, the dam threw a terrific scare into the people in the valley. On the morning of June 10, during a flash flood, a rumor spread through Johnstown that the dam was about to break. This was the first spring in years that there was a head of water of any size behind the mammoth earth embankment, and it was the first of the many springs from then until 1889 that just such rumors would fly from door to door, across back alleys, up and down Main Street, and all along the line between Johnstown and South Fork.

This time the Cambria Iron Company sent two of its men to the lake with instructions to make a critical examination. The dam looked perfectly solid to the Johnstown men, and they returned home with their report in time to make the evening edition of the paper. The fact that they had found the water only two feet from the breast of the dam did not seem to disturb them especially, or the editors of the *Tribune*.

The paper summed up its story as follows: "Several of our citizens who have recently examined the dam state it as their opinion that the embankment is perfectly safe to stand all the pressure that can be brought to bear on it, while others are a little dubious in the matter. We do not consider there is much cause for alarm, as even in the event of the dyke breaking there is plenty of room for the water to spread out before reaching here, and no damage of moment would result."

There it was, in one sentence. In the first place, the dam was probably sound, and even if it did fail not a great deal would happen since the dam was so far away. It was a strange piece of reasoning to say the least, but there it was in the evening paper for everyone, including the alarmists, to read and talk about.

Still, that very night, panic swept through the west end of town, which was the lowest end of town and that part which would have been hardest hit by anything coming down the valley of the Little Conemaugh. People were up through the night "in mortal dread for fear the old Reservoir near South Fork might break," the *Tribune* reported the next day. So apparently the paper could say what it might about "no damage of moment"; people were still unsettled, and especially on nights when the dark and the drenching rain blotted out the landscape and imaginations filled with an ancient terror of death raging out of the mountains.

But nothing happened. Dawn rolled around as usual; the day began. The long shadow of Green Hill slipped back from midtown as the sun climbed into the sky; life went on. And it looked as though the paper and everyone who thought along the same lines were right after all. There was really no cause to get excited.

From then on, practically every time there was high water in Johnstown there would be talk about the dam breaking. One longtime resident was later quoted at length in the newspapers in New York and elsewhere: "We were afraid of that lake . . . No one could see the immense height to which that artificial dam had been built without fearing the tremendous power of the water behind it . . . I doubt if there is a man or a woman in Johnstown who at sometime or other had not feared and spoken of the terrible disaster that might ensue. People wondered and asked why the dam was not strengthened, as it certainly had become weak; but nothing was done, and by and by they talked less and less about it . . ." He also evidently had misgivings about the "tremendous power" of the men who had owned the dam, for he chose to withhold his name.

Others came forth at the same time, that is, *after* May 31, 1889, claiming to have long held doubts about the engineering of the dam, and premonitions of doom for the whole valley of the Conemaugh. Assuredly most of them spoke from deep conviction; but exactly how much widespread, serious, public concern there was, and particularly in the years of the late '80's, is very hard to say.

Certainly there was every reason to have been concerned. The valley from the lake down to Johnstown had sides as steep as a sluice, and there was only one way the water could go if the dam failed. Floods hit the area almost as regularly as spring itself. Johnstown rarely got through a year without water in the streets at least once, and often for several days at a time.

Floods had been a problem from the time of the very earliest settlements in the valley. Lately, for the past ten years or so, they had been getting worse.

The very first flood anyone had bothered to make a record of in Johnstown destroyed a dam. That was in 1808, and it had been only a small dam across the Stony Creek which had been put in as a millrace for one of the first forges. Then there were the so-called Pumpkin Floods of a dozen years later. They hit in the fall and had swept what looked like every pumpkin in Cambria County down into town. In 1847 another little dam on the Stony Creek broke. During the flood of 1875 the Conemaugh rose two feet in a single hour. In 1880 again another dam broke; it had been built by Cambria Iron as a feeder for the mills and was about sixteen feet high, but it was located below town, so no damage was caused. During the next eight years there were seven floods, including three bad ones in 1885, '87, and '88.

The reasons were obvious enough to anyone who took the time to think about the problem, which quite a few were doing by 1889. With the valley crowding up the way it was, the need for lumber and land was growing apace. As a result more and more timber was being stripped off the mountains and near hills, and in Johnstown the river channels were being narrowed to make room for new buildings and, in several places, to make it easier to put bridges across.

Forests not only retain enormous amounts of water in the soil (about 800 tons per acre), but in mountainous country especially, they hold the soil itself, and in winter they hold snow. Where the forests were destroyed, spring thaws and summer thunderstorms would send torrents racing down the mountainsides; and each year the torrents grew worse as the water itself tore away at the soil and what little ground cover there was left. Then, in the valley, where the water was being dumped ever more suddenly, the size of rivers which had to carry it all was being steadily whittled away at by industry and the growing population. So there was always a little less river to handle more runoff, and flash floods were the inevitable result.

Some men in Johnstown, curiously enough, thought that encroaching on the river channels would simply force the water to dig deeper channels. But this was impossible because the river beds were nearly all rock. When the volume of water increased, the rivers only came up, and often very fast. In the 1885 flood the Stony Creek rose three feet in forty-five minutes.

The dam was going to break that year, too, and every year, except one or two, up until 1889. At George Heiser's store, people would come in out of the rain to buy something or just to pass the time in a dry, warm place and nearly always someone said, "Well, this is the day the old dam is going to break." It was becoming something of a local joke. Many years later Victor Heiser would recall, "The townspeople, like those who live in the shadow of Vesuvius, grew calloused to the possibility of danger. 'Sometime,' they thought, 'that dam will give way, but it won't ever happen to us!' "

When there were warnings of trouble up the mountain, very few took them to heart. The dam always held despite the warnings. People got tired of hearing about a disaster that never happened. And after all, was not the dam owned by some of the most awesome men in the country? If there was anything to worry about certainly they would know about it.

The *Tribune* continued to imply that there was no cause for alarm. In 1887 the editors again allowed that a break at South Fork would not greatly affect Johnstown, unless it "occurred in conjunction with a great flood in the Conemaugh Valley which is one of the possibilities not worth worrying about." Readers all through town nodded in agreement. On the afternoon of May 31, 1889, shortly before four, one leading citizen was asked how much higher he thought the water would rise in the valley if the dam let go. His answer was "About two feet."

-3-

It would appear, in fact, that Johnstown's leading citizens had taken little or no intelligent account of the threat the dam posed, were it not for some highly interesting letters that changed hands during the year 1880.

When Benjamin Ruff first began his restoration of the old dam in the fall of '79, the management of the Cambria Iron Company, in the words of its solicitor, Cyrus Elder, became "extremely exercised" over the news. The management at that time was a man by the name of Daniel Johnson Morrell.

Morrell was one of the foremost ironmasters of the age, a ruddy-faced Quaker with gray eyes, who wore his whiskers beneath his jowls, so it appeared he was forever sporting a hair scarf. He looked upon the likes of Carnegie as parvenus in the business, brash, unprincipled upstarts who were not real ironmen at all, but harum-scarum drummers who had jumped into something they knew nothing about just to make a quick fortune. Beside the almost elflike Carnegie or Frick, he looked as though he were of another species. He was under six feet tall, but with his massive, thick shoulders and ample girth (he weighed well over 200 pounds), coming along Main Street he looked every bit the most powerful man in town.

But according to one of his contemporaries, "With all the responsibilities of his position, with all the care and concern of the great works on his hands, he never seemed worried or out of humor. When he left his desk at the close of the day he seemed to be able to shut off all thought of work; and in the midst of other persons' worry and nervousness in the most distressing times, he would lie down and sleep as contentedly as a child."

Aside from running the Iron Company, Morrell presided at most town meetings and was President of the Savings Bank and the First National Bank, the water company and the gas company. He had served two terms in Congress and was still a powerful voice in the Republican Party. For many years he was the President of the American Iron and Steel Association, an organization which did as much as any to protect the protective tariff.

He lived on Main Street in the finest house in Johnstown, a tall brick house with a mansard roof, painted white and set among gardens and shade trees on a lawn that took up a full city block. He had the only greenhouses in town, a full-time gardener, and all his property was enclosed with an ornamental iron fence. Children used to gather by the fence after school, hoping for a chance to look at him. "Whatever Mr. Morrell wants, well that's it," they heard at home. He was the king of Johnstown.

Morrell had been born in Maine, in 1821 (which made him fourteen years senior to Carnegie), but grew up in Philadelphia and started out clerking in a mercantile store. He had moved to Johnstown in the 1850's when the Philadelphia financial backers of the then floundering Cambria Iron Company sent him to see what might be done to keep the works from going bankrupt.

Backwoods iron forges had been in operation in Cambria County for fifty years and more. With plenty of ore, limestone, and coal in the locale, the prospects for turning the Conemaugh Valley into an iron center of some real consequence looked extremely bright. But until Morrell came to town the industry had been beset by repeated failures. Morrell, however, succeeded handsomely. Knowing nothing about the iron business, he reorganized the company, and despite fires and financial panic, he kept his nerve, maintaining to the Philadelphia money men that the works would one day prosper.

By the start of the Civil War the Cambria Iron Company was the biggest iron-producing center in the country. In addition, Morrell had encouraged some rather primitive and haphazard research into a new pneumatic process for making steel which contributed substantially to dramatic changes in the iron business and, for that matter, in the whole character and growth of the country.

In 1856 a man named William Kelly, a Pittsburgher by birth, moved from Eddyville, Kentucky, to Johnstown to set up in one corner of the Cambria yard some experimental apparatus which he assembled from scrap-heap parts and pieces. Kelly was in Johnstown off and on for the next three years. He became known among the millworkers as "The Irish Crank," and not without justification. His attempts to "refine" molten iron for the rolling mill by blowing air into it had resulted in repeated failures and at least one serious fire, which became known as "Kelly's Fireworks." But later on, in 1862, he came back to try again, this time with an egg-shaped "converter" made abroad, and the accepted story is that he had better luck. Kelly would later be credited with having built the converter himself and with developing at Johnstown something very close to what became known as the Bessemer process, a technique for converting iron into steel at far less cost and in considerably larger quantities than had been possible before.

Henry Bessemer, a brilliant English chemist, had devised just such a process at about the time Kelly first arrived at the Cambria works, and, deservedly enough, got nearly all of the credit. The Bessemer converter used a blast of air directed through molten iron to oxidize, or burn off, most of the carbon impurities in the metal to make steel. Previous steelmaking techniques required weeks, even months. The Bessemer process could produce good-quality steel in less than one hour.

It was one of the important technological innovations of all time, and Morrell was among the first to recognize just what its impact might be. He financed Kelly's erratic pioneering in the technique for close to five years and after the war invested heavily in new Bessemer equipment. In the late '60's and '70's Johnstown was the liveliest steel center in the country, with the most inventive minds in the industry gathering there—the Fritz brothers, George and John, Bill Jones, and the brilliant and energetic Alexander Holley.

Moreover, Morrell had Cambria Iron do something no other steel company experimenting with the Bessemer process dared try, and something that was to prove immensely beneficial to Andrew Carnegie. He used only American workers, training Pennsylvania farm boys to understand and master the new technology, while everyone else in the business was importing English workers already familiar with it. At first there were months of costly setbacks and disappointments in Johnstown, but the results in the long run proved Morrell right.

In 1867, from ingots made at Steelton, the first Bessemer rails to be rolled on order in the United States came out of the Cambria mill. By 1871 Morrell had one of the first really big Bessemer plants in operation, and for the next five years Cambria would be the largest producer in the country, if not the world.

The war had brought flush times and dazzling increases in iron production capacities. But now the age of cheap steel was on. By the time of the late 1880's, Cambria Iron had some 7,000 men on the payroll. The works consisted of the Johnstown furnaces Numbers 1, 2, 3, and 4 in one plant, with stacks seventy-five feet high and sixteen feet in diameter at the base. Blast furnaces Numbers 5 and 6 were in a second plant. The hulking Bessemer plant was the main building. Then there was a huge open-hearth building, a rolling mill that was nearly 2,000 feet long, a bolt-and-nut works, and an axle shop.

Added to that the company owned and operated its own coal mines, coke ovens, and railroads. It was the largest landowner by far in the county, having bought up thousands of acres around Johnstown, coal holdings primarily, which were, in many places, used as tremendous farms where nothing but hay was grown to feed the animals used in the mills and mines. The company also owned some 700 frame houses which it rented to its workers, a big department store, and the Gautier Steel Company, a subsidiary, where Cambria Link Barbed Wire was made.

To all intents and purposes, Johnstown, in other words, was a company town and an important one at that. And appropriately enough the company ran the place with an iron hand. Labor unions were not to be tolerated, nor were employees who dared even to talk such treason.

For example, Rule Number 9 of the plant regulations published in 1874 stated: "Any person or persons known to belong to any secret association or open combination whose aim is to control wages or stop the works or any part thereof shall be promptly and finally discharged. Persons not satisfied with their work or their wages can leave honorably by giving the required notice . . ."

The Cambria Iron Company, which meant Mr. Daniel J. Morrell, left no doubts as to where it stood on such matters. So there were no unions in the mill, and inside the high, green fence that surrounded it, work went on around the clock, around the calendar, without any trouble from the help.

It would be mistaken, however, to imagine Cambria Iron as an entirely overbearing or inhuman organization, grinding down its employees. By the standards of the day, it was quite progressive and looked out for the welfare of its people and the town with uncommon paternalism.

In his first speech in Congress, Morrell had said, "The American workingman must live in a house, not a hut; he must wear decent clothes and eat wholesome and nourishing food. He is an integral part of the municipality, the State, and the Nation; subject to no fetters of class or caste; neither pauper, nor peasant, nor serf, but a free American citizen." Judged by the standards of his time, he was almost as good as his word.

In one of its plants the Iron Company maintained the eight-hour day, a practice that had been tried and abandoned by every other steel company, which meant, as one of the trade-union newspapers pointed out, that the only eight-hour mill left in the country was a nonunion mill.

The town hospital was built by the company and anyone injured on the job received free treatment there. It was the company also which had established the library and a night school where its employees could learn elementary science, mechanical drawing, and engineering. At the company store, Wood, Morrell & Company, which advertised itself as "The Most Extensive and Best Appointed Establishment in its Class in the United States," prices were quite reasonable. At the time the South Fork Fishing and Hunting Club was organized, Cambria Iron had somewhere in the neighborhood of $50 million invested in Johnstown and along the valley. So Morrell, very understandably, had special interest in what the Pittsburgh men were up to. That he held no special good feelings toward some of the clubmen also seems likely.

When Carnegie, who had once stoutly proclaimed that "pioneering don't pay," decided the time was right to get into the Bessemer steel business and Carnegie, Phipps & Company built the Edgar Thomson works at Braddock in the early 1870's, he had raided Cambria Iron of its best workers. Among these men, most of whom had been working with pneumatic conversion techniques since Kelly's days in Johnstown, was the tough, gifted little Welshman, Bill Jones.

Captain Bill Jones, as he was known, had been with Morrell for sixteen years. He had acquired a vast knowledge of the new process and had built a tremendous following among the men at Cambria Iron, largely for his robust, freewheeling willingness not to do things according to the rules if the rules did not suit him. It was the sort of reputation Carnegie took a special interest in. Carnegie never did learn much about steelmaking, but he had a gift for finding men who did, and if they were somewhat unorthodox, so much the better.

When the manager of the Cambria works, George Fritz, died and Morrell had to pick a man to replace him, he turned Jones down for the job, giving it instead to another Jones who seemed a steadier sort. At which point Carnegie immediately moved in and offered Bill Jones a two-dollar-a-day job at Braddock. Jones not only accepted but took a number of his "high-class graduates" along with him. Within one year, by 1876, the Edgar Thomson mill moved ahead of Cambria in production. Carnegie, on a mountaintop in Italy, literally danced with joy on hearing the news. And as Edgar Thomson continued to break every other production record, it was Bill Jones, not Carnegie or his associates, who got the credit from anyone who knew anything about steelmaking. Jones's papers on production techniques were read before learned societies in Europe. In Pittsburgh he became so important to the Carnegie empire that Carnegie decided to make him a partner, an offer which Jones flatly refused, feeling that he would lose his influence with the men if he ever so openly joined forces with management.

"Just pay me one hell of a salary," Jones said to Carnegie.

To which Carnegie shot back, "All right, Captain, the salary of the President of the United States is yours."

"That's the talk," said Jones. The salary was $25,000 a year.

So now, if the Pittsburgh crowd was about to go tampering with the South Fork dam, Daniel J. Morrell wanted to know more about their doings than could be gained from mere hearsay. There was too much at stake to go on their word alone. He had no intention of stopping them, as he made clear later on. He had never been one to stand in the way of progress. He had welcomed innovations throughout his working life, and it seems he never objected to this one on principle. He only wanted to be satisfied that the work was being properly managed. He had seen enough explosions and fires at the mill to have a fair idea of the violent consequences of bungled innovation. He had also had some experience with dams, having personally supervised the installation of several small ones put in near town by the water company.

In November of 1880 he sent John Fulton to look over the

job. Fulton was an engineer by training and profession, but he was also the next in line to succeed Morrell as head of the works. Morrell, in other words, was not just sending any ordinary employee to South Fork.

A lean Ulster County Irishman, Fulton wore his beard close-cropped and had a fix to his mouth like General Grant's. He had wonderfully heavy eyebrows and a resolute gaze that gave him the look of an Old Testament prophet. He was a man to reckon with, one of Johnstown's most ardent temperance leaders and a pillar at the Presbyterian Church, where he taught Sunday school and would be long remembered for closing his Bible classes with the most interminable prayers ever uttered by man.

Fulton had made his reputation as a mining engineer and geologist before joining Morrell at Cambria Iron. That there was anyone in Johnstown better qualified to pass judgment on the dam is doubtful. Nor was there any man, save Morrell himself, who was less likely to be dazzled or cowed in any way by the representatives of the club, a factor which Morrell must have taken into account.

Fulton was met at South Fork by two club members, Colonel E. J. Unger and C. A. Carpenter, as well as some of the contractors who had worked on the dam. His report was filed in a letter to Morrell dated November 26. The letter began by stating that he had gone as requested to inspect the dam now owned by the "Sportsmen's Association of Western Pennsylvania." (The correct name of the club was evidently still unknown to the Cambria management.) He then said that he did not think the repairs were done in "a careful and substantial manner, or with the care demanded in a large structure of this kind." He stated that he believed the dam's weight was sufficient to hold back the water, but that he had grave misgivings about other aspects of the dam:

> There appear to me two serious elements of danger in this dam. First, the want of a discharge pipe to reduce or take the water out of the dam for needed repairs. Second, the unsubstantial method of repair, leaving a large leak, which appears to be cutting the new embankment.
>
> As the water cannot be lowered, the difficulty arises of reaching the source of the present destructive leaks. At present

there is forty feet of water in the dam, when the full head of 60 feet is reached, it appears to me to be only a question of time until the former cutting is repeated. Should this break be made during a season of flood, it is evident that considerable damage would ensue along the line of the Conemaugh.

It is impossible to estimate how disastrous this flood would be, as its force would depend on the size of the breach in the dam with proportional rapidity of discharge.

The stability of the dam can only be assured by a thorough overhauling of the present lining on the upper slopes, and the construction of an ample discharge pipe to reduce or remove the water to make necessary repairs.

Morrell promptly sent the report to Ruff, who responded on December 2.

Ruff was not much impressed by Fulton's findings. He pointed out to Morrell that Fulton did not have the correct name of the club, and told Morrell what that name was. He said there was no leak such as Fulton claimed and that Fulton's figures on the comparative weights of the water and the dam were off, since Fulton had overestimated how much water was in the lake. The tone was one of obvious impatience and suggested not very subtly that Morrell would do well to hire himself a more competent man. He ended the letter by saying:

> We consider his conclusions as to our only safe course of no more value than his other assertions. . . . you and your people are in no danger from our enterprise.
>
> Very respectfully,
> B. F. Ruff, President

Ruff, quite clearly, was not in the least interested in continuing the discussion. The club had managed nicely to keep its affairs private until then, and the idea of any prolonged or possibly complicated negotiations with the Cambria Iron Company had small appeal.

Morrell, however, was unwilling to let it go at that. On December 22 he answered Ruff's letter. After a few opening courtesies, he got to the heart of the issue:

. . . I note your criticism of Mr. Fulton's former report, and judge that in some of his statements he may have been in error, but think that his conclusions in the main were correct.

We do not wish to put any obstruction in the way of your accomplishing your object in the reconstruction of this dam; but we must protest against the erection of a dam at that place, that will be a perpetual menace to the lives and property of those residing in this upper valley of the Conemaugh, from its insecure construction. In my judgment there should have been provided some means by which the water would be let out of the dam in case of trouble, and I think you will find it necessary to provide an outlet pipe or gate before any engineer could pronounce the job a safe one. If this dam could be securely reconstructed with a safe means of driving off the water in case any weakness manifests itself, I would regard the accomplishment of this work as a very desirable one, and if some arrangement could be made with your association by which the store of water in this reservoir could be used in time of drouth in the mountains, this Company would be willing to cooperate with you in the work, and would contribute liberally toward making the dam absolutely safe.

Morrell, in short, was suggesting exactly what Fulton had urged: give the dam a major overhaul and install a discharge system of some sort. At the same time, he was making it plain that Cambria Iron considered the present job shoddy enough, the situation critical enough, to be willing to help foot the bill to set things right.

The offer was declined. The matter was dropped—almost.

Morrell felt that just to be on the safe side it might be a good idea to have an inside view of doings at the lake. So he decided to join the South Fork Fishing and Hunting Club, and evidently the Pittsburgh men had no objections. Morrell therefore purchased two memberships in his own name.

It was not for another nine years that engineers from many parts of the country came to the site of the dam to study what had gone wrong. Fulton's findings appeared to have been correct. But there were four other changes in the dam which Ruff and his men had made which Fulton had not noticed, and these were as crucial to what finally happened as the faults Fulton cited.

To begin with, in order to provide room for a road across the breast, the height of the dam had been lowered from one to three feet. This would give enough width for two carriages crossing the dam to pass each other comfortably. But it also meant that the capacity of the spillway had been reduced, for now the bottom of the spillway was not ten or eleven feet lower than the crest of the dam, but perhaps only seven or eight feet. This was a very significant change, since it meant that a rising lake would start to go over the top of the dam that much sooner.

Then, too, a screen of iron rods, each about half an inch in diameter, had been put across the spillway to prevent the fish from going over and down into South Fork Creek. The screen was set between the heavy posts which supported the wooden bridge over the spillway. Under normal conditions the combination of posts and screens decreased the spillway capacity only slightly, but they had the potential of decreasing it a great deal should the screens become clogged with debris.

The third change was probably the most important of all. The dam sagged slightly in the middle, where the old break had been. Exactly how bad the sag was no one was able to say later for certain. It may have been only a foot or two, but according to one study, the crest at the center may have been as much as four feet lower than the ends. The center was where the dam should have been highest and strongest, so in the event that water ever did start over the top, the pressure would be at the ends rather than at the middle. Now the reverse was the case.

To have seen the sag with a naked eye, and particularly an untrained eye, would have been next to impossible. It is conceivable therefore that it went unnoticed by Ruff and the men who did the reconstruction work. Fulton took no note of it apparently; whether it would have been observed and corrected had experienced engineers been responsible for the reconstruction is a question no one can answer.

What it meant in practical terms was that the depth of the spillway was now only about four feet lower than the top of the dam at its center. In other words, if more than four feet of water were going over the spillway, then the lake would start running over the top of the dam at the center where the pressure against it was the greatest.

The fourth change was unnoticed by Fulton because it had not as yet taken place when he made his inspection. The water then, as he says, was only forty feet deep, which is about the depth it had been kept at during the old days before the first break in 1862. The club, however, brought the level of the lake up to where it was nearly brim full, meaning that the depth ran to sixty-five feet or thereabouts. In spring it sometimes rose even higher. With the lake that full, it was not beyond reason to imagine serious trouble in the event of a severe storm.

But, as both Fulton and Morrell had made abundantly clear, with the discharge pipes gone, the club was faced with the unfortunate position of not being able to lower the level of the lake, ever, at any time, even if that were its expressed wish.

The water that high at the dam also meant that the over-all size of the lake was increased. The lake backed up well beyond where it had been in the old days, which lead to the widespread misconception, still current today in and around Johnstown, that the club had actually raised the height of the dam from what it had been.

How satisfied Morrell was after the business of the letters was over and done with is not known. For when the sun went down behind Laurel Hill on Monday, August 24, 1885, Daniel J. Morrell was in his grave at Sandy Vale. He had been ill for several years, having suffered what appears to have been an advanced case of arteriosclerosis. He had gone into a steady mental decline not long after he took out his membership in the South Fork fishing club. In 1884 he had given up all his various civic responsibilities and retired from business. After that, it seems, senility closed in hard and fast. He was seen almost never, "lost in mental darkness," as one account put it years later. When he died, "calmly and peacefully" at eight in the morning on Thursday, August 20, 1885, he was sixty-four years old.

On Sunday thousands of mourners queued up along the south side of Main Street to go through the iron gates, up the long front walk, and into the big house to view the remains. For three hours the doors were open and a steady procession filed through.

The next day, from noon until five, the whole town was shut down. The procession that marched out to the cemetery was as fine a display of the town's manhood as anyone had ever seen. Ahead of the hearse tramped men from the Cambria mines and railroads, the rolling mills and blast furnaces, row on row, like an army, followed by the merchants and professional men, the police, the city fathers, men of every sort who worked for or did business with or depended on the Cambria Iron Company, which meant just about everybody. The only sound was the steady beat of their heavy boots and shoes on the cobblestones.

After the hearse came the special carriages for the mourners. Bill Kelly and his wife were there; so was Captain Bill Jones, and a Cleveland steel man and family friend named Marcus Alonzo Hanna.

There never was a bigger or better funeral in Johnstown.

Two years later, on March 29, 1887, the day a wagonload of fruit trees arrived at his cottage on Lake Conemaugh, Benjamin F. Ruff died suddenly in a hotel in Pittsburgh. The cause of death according to the papers was a carbuncle on the neck.

III

"There's a man came from the lake."

-1-

The hard, cold rain that had started coming down the night before had eased off considerably by the morning of Friday, May 31. But a thick mist hung in the valley like brushwood smoke and overhead the sky was very dark.

Even before night had ended there had been signs of trouble. At five o'clock a landslide had caved in the stable at Kress's brewery, and anyone who was awake then could hear the rivers. By six everyone who was up and about knew that Johnstown was in for a bad time. The rivers were rising at better than a foot an hour. They were a threatening yellow-brown color and already full of logs and big pieces of lumber that went bounding along as though competing in some sort of frantic race.

When the seven o'clock shift arrived at the Cambria mills, the men were soon told to go home and look after their families. By ten there was water in most cellars in the lower part of town. School had been let out, and children were splashing about in the streets with wooden boxes, boards, anything that would make a boat.

One of the most distinguished residents of the lower part of

79

town was the Honorable W. Horace Rose, Esquire, former cavalry officer, former state legislator, former district attorney of Cambria County, a founder of Johnstown's Literary Society, father of five, respected and successful attorney at law. Rose was a Democrat with a large following among Republicans as well as his own party, and including those Republicans who ran the town and Cambria Iron. He was an expert horseman, slender, erect, and full-bearded, with strong blue eyes and a soft voice which he seldom ever raised.

He had been born in a log house that had stood at the corner of Vine and Market. At thirteen he had been orphaned when both parents died of cholera within the same hour and had been on his own ever since, first as a bound boy in a tannery, later as a carpenter. When he was nineteen, John Linton, Johnstown's leading lawyer, took him into his office to "read law." Not long after he opened his own office, which he built himself, and got married to Maggie Ramsey of Johnstown. Then came the war, during which he was wounded, captured, and released in time to take part in General Sheridan's Shenandoah Valley campaign.

The house Horace Rose and his family lived in was downtown, on lower Main Street. He had witnessed nearly every one of Johnstown's floods over the years, and when he heard that the rivers this particular morning were both coming up rapidly at the same time, something they had not done before, he decided to go out after breakfast and see how things were going.

By the time he and his two youngest sons, Forest and Percy, had finished hitching the team, there was water on the stable floor. Rose took care to use his second-best harness and had one of the boys drive their cow to the hillside, expecting to bring her back down again in a few hours, when the water subsided. Then they climbed into an open wagon and headed down Main, with Rose driving.

His intention was to pick up anyone toward the river who wished to be evacuated. But by now, which was somewhere near nine, the water that far downtown was too deep to get through safely. So he wheeled around and headed back up Main, going as far as Bedford, where he paused to pass the time with his old friend Charles Zimmerman, the livery-stable owner. For a few moments they watched another cow being led to higher ground.

"Charlie," Rose said, "you and I have scored fifty years, and this is the first time we ever saw a cow drink Stony Creek river water on Main Street."

"That's so," Zimmerman agreed. "But the water two years ago was higher."

About then the rain started coming down again as hard as it had during the night, heavy and wind-driven. Rose stopped off long enough to buy his sons rubber raincoats, then proceeded over to the end of Franklin to his office, which was less than a hundred feet from the roaring Stony Creek. For another half-hour he and the boys set about placing his papers and other things well above the flood line of 1887, which was about a foot from the floor. Then they started for home, stopping once more on the way to talk to another old friend, John Dibert, the banker, who was also their next-door neighbor.

The situation for property holders in the lower part of town was growing serious, Rose and Dibert agreed. This business of flooded cellars every spring had to be corrected. The solution, as they saw it, was to call a meeting to protest the way the Cambria Iron Company had been filling in the riverbanks next to the mills below town. They recalled that town ordinances had fixed the width of the Stony Creek at 175 feet and the Little Conemaugh at 110 feet. This meant that the combined width of the two was 285 feet; but the Conemaugh, which had to carry all the water from both of them, was now less than 200 feet wide near the mills. Obviously, the rivers were bound to back up when flash floods hit, and obviously the Cambria Iron Company would have to restore the riverbed to its original width. With that settled, the two friends parted.

Rose went directly home, where he found the water now so deep that he was unable to get near his front door. He sent young Forest with the team to a nearby hillside while he and Percy assembled a makeshift raft and floated over to the back porch. Once inside, like nearly everyone else in town, they busied themselves taking up carpets and furniture. Rose also "marked with sadness" that the slowly rising water "with its muddy freight" had already ruined his new wallpaper.

Then everyone moved upstairs where the morning took on the

air of a family picnic. Forest had been unable to get back to the house after leaving the horses but had signaled from a window across the street that he was high and dry with the Fishers.

Horace Rose called back and forth to Squire Fisher and joked about their troubles while his wife and the others got a fire going in the grate and made some coffee. After a bit Rose got his rifle, went up to the attic, propped himself in a window, and whiled away the time shooting at rats struggling along the wall of a stable in the adjoining lot.

And so the morning passed on into afternoon; there was nothing much to do but wait it out and make the best of what, after all, was not such an unpleasant situation.

There were hundreds of other families, however, who had seen enough. They began moving out, wading through the streets with bundles of clothing and food precariously balanced on rude rafts, or jammed into half-submerged spring wagons. Here and there a lone rowboat pushed up to a front porch or window ledge to make a clumsy, noisy rescue of women and grandfathers, dogs, cats, and children.

Some people were simply heading for higher ground, without any particular place in mind; others were going to the homes of friends or relatives where they hoped there might not be quite so much water, and where, come nightfall, there might be electricity and a dry kitchen.

A few families went over to the big hotels in the center of town, thinking they would be the safest places of all to ride out the storm. Quite a good many, wherever their destination, went a little sheepishly, dreading the looks and the kidding they would get when they came back home again.

The water by now, from one end of town to the other, was anywhere from two to ten feet deep. It was already higher than the '87 flood, making it, by noon at least, Johnstown's worst flood on record. The Gautier works had closed down at ten, when Fred Krebs, the manager, was reminded by one of the men that the huge barbed-wire plant stood on fill that had been dumped into the old canal basin, and that once upon a time there had been four feet of water right where they were standing. At eleven, or soon after, a log boom burst up the Stony Creek and sent a wild rush of logs stampeding through the valley until they crashed into the stone bridge below town and jammed in among the massive arches.

Not very long after that the Stony Creek ripped out the Poplar Street Bridge; then, within the hour, the Cambria City Bridge went. At St. John's Catholic Church, which stood far uptown at Jackson and Locust, and so, presumably, well beyond reach of spring floods, the water was so deep that the funeral of Mrs. Mary McNally had to be postponed midway through the service and the casket left in the church.

Worst of all, and unlike any other flood in Johnstown's history, there had been a tragic death. A teamster named Joseph Ross, a father of four children, had been drowned when he fell into a flooded excavation while helping evacuate a stranded family.

Along Main Street, shopkeepers were working feverishly to move their goods out of reach of the water. In his second-floor office overlooking Franklin and Main, George T. Swank, the cantankerous editor and proprietor of the *Tribune*, began working on what he planned to be a running log of the day's events, with the intention of publishing it in the next edition, whenever that might be.

"As we write at noon," he put down, "Johnstown is again under water, and all about us the tide is rising. Wagons for hours have been passing along the streets carrying people from submerged points to places of safety . . . From seven o'clock on the water rose. People who were glad they 'didn't live downtown' began to wish they didn't live in town at all. On the water crept, and on, up one street and out the other . . . Eighteen inches an hour the Stony Creek rose for a time, and the Conemaugh about as rapidly."

On the street below his window the current, coming across from the Stony Creek, was rushing by at an estimated six miles an hour.

Across Main, and three doors down Franklin, the Reverend H. L. Chapman was having a slightly unnerving day.

After an early breakfast he had retired to his study to work on his sermon for Sunday. The text he had selected was "But man dieth, and wasteth away: yea, man giveth up the ghost, and where is he?" He had barely begun when he was interrupted by the doorbell. Opening the front door, he found his wife's cousin, Mrs. A. D. Brinker, standing on the porch looking terribly frightened.

She had crossed the park from her home on the other side. She asked Chapman if he had heard about the high water downtown. He said he had not and that he did not think there would be much of a flood.

"Johnstown is going to be destroyed today," she said, and then told him that the reservoir would break and all would be swept away.

Chapman was so incredulous he almost laughed in her face, and would have had she not looked so pitifully terror-stricken. It was also not the first time Mrs. Brinker had made just such a forecast.

"Well, Sister Brinker, you have been fearing this for years," Chapman said with patience, "and it has never yet happened, and I don't think there is much danger."

He invited her to step in and stay with them until after the flood had passed, an invitation she gladly accepted, saying that her husband had insisted on staying home, to "hold the fort" as he had put it.

Later, a young student friend named Parker dropped by to see if the Chapmans needed help moving their furniture, but the Reverend told him he expected no trouble, as the new parsonage had a higher foundation than other houses. But the young man stayed on nonetheless.

About noon Chapman happened to look out the window long enough to see one of the town's most dignified figures standing in the street in water up to his waist. It was Cyrus Elder, who along with being chief counsel for the Iron Company was now Johnstown's one and only member of the South Fork Fishing and Hunting Club, having acquired the Morrell memberships at the time of the old Quaker's death.

Chapman was puzzled by the whole thing and hurried onto the porch to see what he could do. Elder, though a most solid citizen, seems also to have had a sense of humor.

"Doctor," he called, "have you any fishing tackle?"

Chapman answered that he thought he had.

"Well," said Elder, "I was in a skiff and it upset and left me here, and I am waiting for a man who has gone after a horse, to take me out, and I might put in my time fishing."

When the man and horse returned, the hefty Elder, try as he would, was unable to get up on the animal's slippery back. So the rider went off again and returned next with a wagon. This time they had better luck and started off down Main toward Elder's home on Walnut Street but had to turn back and head uptown for Elder's brother's house; all of which was duly noted with amusement from the window of the *Tribune* by George Swank, who was also Elder's brother-in-law.

Elder had arrived back in Johnstown from Chicago just that morning and had been trying for hours to get home to his wife and daughter. From the station platform he had been able to look right across at his house, where, on the front porch, the two women were waving their handkerchiefs at him. They had gestured back and forth about the water and how he might get home, but from then on he had made little progress.

At midday the Chapmans and their guests sat down to dinner, but Mrs. Brinker was still too unstrung to eat anything. Dinner over, they moved to the study, where they sat quietly chatting beside the gas fire. But the Reverend soon grew impatient with the comings and goings of Lizzie Swing, the Chapmans German servant girl, who kept tramping past the study door on her way from the cellar to the attic with armloads of food.

"Why does she do that?" Chapman asked. He was genuinely puzzled by the girl's obvious state of nerves, since she understood almost no English. The Reverend was having trouble maintaining an intelligent sense of domestic calm.

The train which returned Cyrus Elder to Johnstown had left Chicago the day before at three in the afternoon. It was one of two sections of the *Day Express*, which had pulled out of Pittsburgh that morning, on time, at 8:10. The two trains arrived at Johnstown about 10:15, and again on time, but were held there for a half-hour or so. The eastbound track on up the valley had washed out. Not until a local mail train came through were the two sections given orders to follow it to East Conemaugh, running east on the westbound track.

During the wait at Johnstown, passengers on board watched in fascination the struggles of the flooded city. They waved back and forth to families hanging out of upstairs windows. Some got out for a few minutes and joined the crowds on the station platform and along the near enbankment to watch the railroad crew that was trying to dislodge the logs and drift from the stone bridge. When the trains began moving again, very slowly, around the blind corner of Prospect Hill and on to the East Conemaugh yards two miles ahead, the passengers could see the ugly yellow-brown surge of the Little Conemaugh to their right, now very near their rain-streaked windows. More and more debris swept by and telegraph poles swayed precariously in the strong wind.

The run to East Conemaugh took about ten minutes, with the mail train in the lead, followed by the first and second *Day Express*, in that order. At East Conemaugh all trains were stopped and held for further orders. There was trouble up the line at Lilly. Bear Run had risen more than six feet, burst over its banks, and washed out a quarter of a mile of track. Nothing could move east or west. Two earlier eastbound trains, the *Chicago Limited* and a freight from Derry, had gotten as far as South Fork, where now they too were being held.

The first section of the *Day Express* was made up of seven cars—five coaches, a baggage car, and one Pullman. On board were some 90 passengers plus crew. The second section had three sleepers, one baggage and two mail cars, and perhaps 50 passengers and crew. The mail train had only three cars, one for the mail, plus two coaches. Most of its passengers were members of the *Night Off* company, who were on their way to Altoona for their next performance.

East Conemaugh was the main marshaling yard for Johnstown. Eastbound trains picked up their "helpers" there, the extra engines for the climb over the mountain. There was a huge sixteen-stall roundhouse, water towers, four main tracks, sidings, sheds, repair shops, coal tipples, dozens of locomotives, rolling stock of every description. So that the trains now detained there in the driving rain were only part of a whole concentration of equipment set along a broad flat where the Little Conemaugh makes a sweeping curve between the hills.

For the people of the little town set just back from the yards, the morning was turning into a fine show. The river was still within its banks but rising fast. There was every sign that the wooden bridge above the station was about to wash out, and rain or no rain, most of the town had gathered along the riverbank to see it happen.

Sometime between noon and one o'clock a telegraph message came into the East Conemaugh dispatcher's tower from the next tower up the valley to the east. The message was directed to the yardmaster at East Conemaugh, J. C. Walkinshaw, and to the head of the entire division, Mr. Robert Pitcairn, at Pittsburgh. No one would later recall at exactly what moment the message arrived, and numerous people who should have seen it later claimed that they never did. Nor was there ever agreement on its precise wording, but the consensus was that the message said something close to this:

SOUTH FORK DAM IS LIABLE TO BREAK: NOTIFY THE PEOPLE OF JOHNSTOWN TO PREPARE FOR THE WORST.

It was signed simply "Operator."

At Johnstown the message was received at the telegraph office at the depot only a few minutes later. The freight agent, Frank Deckert, was told it had come in; he glanced at it, but he did not stop to read it. As he said later, he knew that "it was in regard to the dam; that there was some danger of it breaking." But it created no alarm in his mind. He had heard such warnings before. When he passed the word along to the few people who happened to be about the station, their response was the same as his, with only one or two exceptions. Two men who were shown the message by Charles Moore, the assistant ticket agent, read it and laughed out loud.

Deckert made no further effort to spread the warning. He did not move his family from their home just down from the station, nor did he bother to send the message over to the central part of town.

–2–

No one on the mountain could remember there ever being a night like it.

John Lovett, who was seventy-one and had a sawmill on South

Fork Creek a quarter of a mile from the head of Lake Conemaugh, said it was the hardest rain he had ever heard. He could not see it, he said, but he could hear it all right, and the creeks got so vicious they carried off logs that had been on his place for forty years. William Hank and Sam Peblin, who had farms farther up the mountain, at the headwaters of South Fork Creek, said much the same thing. Sylvester Reynolds, another farmer, reported that Otto Run, which feeds into Yellow Run, was running four feet deep, compared to its normal depth of two inches. F. N. George, Justice of the Peace at Lilly, said he had never known a cloudburst like it in fifty years. At Wilmore, H. W. Plotner, a druggist who was nearly seventy, said he could recall no worse storm. Dan Sipe, who owned the flour mill on the Little Conemaugh at Summerhill, Sheriff George Stineman, the coal operator at South Fork, and Mrs. Leap, who kept the general store at Bens Creek, all agreed it was the mightiest downpour and the highest water ever in their memories.

There were also "weird and unnatural occurrences" reported. One family by the name of Heidenfelter later described how they had been suddenly awakened and badly frightened by a "rumbling, roaring sound" that seemed to come from some indefinable object not far from their house. It was then followed by a terrific downpour, which, according to Mrs. Heidenfelter, sounded as if a gigantic tank had opened at the bottom and all the water dumped out at once.

"Indeed I thought the last day had come," she later told a newspaper reporter. "I never heard anything like it in my life. I wanted my husband to get up and see what the matter was, but it was dark and he could have done no good. In the morning, as soon as we could see, the fields were covered with water four or five feet deep. . . . People say the noise we heard was a waterspout, but I've never seen one and don't know how they act."

Apparently the storm did tear big holes in the ground near the Heidenfelter farm, and other families close by the lake reported hearing sounds much like thunder but which they were certain were not thunder.

In any case it was a wild night on the mountain, and when morning came virtually every farm in the area had swamped cellars and pastures. Freshly plowed fields were sliced through with gullies that carried water as much as three feet deep. Acres of winter wheat and corn planted a few weeks earlier had been washed away. Every backwoods road had turned into a creek; every little mountain spring, run, creek, and stream was on a rampage. The earth could not absorb any more water.

It was about six thirty that morning when young John Parke awoke in his high-ceilinged room upstairs at the clubhouse on the shore of Lake Conemaugh. He had awakened once before, about an hour earlier, and had heard the rain hammering against the big frame building, but thinking nothing of it, had dropped off to sleep again. Now, outside his window, there was little to be seen but a heavy, white mist that had closed down over the trees and water.

Parke dressed quickly, went downstairs, crossed through the main living room, out the porch door into the cold morning, where, for the first time, he heard what he would later describe as a "terrible roaring as of a cataract" coming from the head of the lake to the south. He also noticed that during the night the lake had risen what looked to be perhaps two feet. Yesterday the water had been at its usual level, which, he reckoned, was about four to six feet below the crest of the dam. Now, it might be no more than two or three feet from the crest. What he could not tell was how much water was still coming in, and that he knew would be the crucial factor for the next several hours. But the sound from the head of the lake was far from encouraging.

He went inside again, had breakfast, then, along with a young workman who had been helping on the sewer project, he got hold of a rowboat and started off to have a look at the incoming creeks.

"I found that the upper one-quarter of the lake was thickly covered with debris, logs, slabs from sawmill, plank, etc.," he wrote afterward, "but this matter was scarcely moving on the lake, and what movement there was, carried it into an arm or eddy in the lake, caused by the force of the two streams flowing in and forming a stream for a long distance out into the lake."

As he and his companion neared the far end, he was astonished to discover that they were rowing over the top of a four-strand barbed-wire fence which stood well back from the normal shore line. Then, rowing against the strong current, they proceeded to cover another hundred yards or more across what was normally a cow pasture. They passed by the place where Muddy Run emptied into the lake and went on to South Fork Creek, which Parke described as "a perfect torrent, sweeping through the woods in the most direct course, scarcely following its natural bed, and stripping branches and leaves from the trees five and six feet from the ground."

The two of them pulled their boat onto what seemed the driest spot in sight and started up along the creek by foot. For half a mile the woods boiled with water. The trees dripped water, their drenched trunks black against the mist. The very air itself seemed better than half water.

When they returned for their boat, they found that the lake had come up enough in that short time to set it slightly adrift. From there they struck out straight for the clubhouse. From what he had seen, Parke knew that the situation at the dam must be growing very serious and that an appreciable letup in the volume of water pouring into the lake was most unlikely.

As near as he could tell, the lake was rising about an inch every ten minutes. If this were so, it would be only a matter of hours until the water started over the top of the dam, unless something could be done to release more water than the spillway was handling.

At the clubhouse Parke was told that he was needed at the dam immediately. He went to the stable for his horse and within minutes was galloping off through the cold rain.

There were close to fifty people at the dam when he came riding out of the woods. There was a clump of bystanders, South Fork men and boys mostly, under the trees over at the far side, next to the spillway. Along the road that crossed over the dam itself, a dozen or so of the Italian sewer diggers were working with picks and shovels, trying, without much success, to throw up a small ridge of earth to heighten the dam. Bill Showers, Colonel Unger's hired man, was also making little progress with a horse and plow. Despite all the rain, the road was so hard packed that thus far they had managed to make only a slight strip of loose earth across the center of the dam hardly more than a foot high.

At the center of the dam the water level was only two feet or so from the top.

At the west end another ten or twelve men were trying to cut a new spillway through the tough shale of the hillside but were able to dig down no more than about knee-deep, and the width of their trench was only two feet or so.

Also among the onlookers were several of the clubmen who had come up from Pittsburgh for the Memorial Day weekend. But the man who was directing things, and deciding what ought to be done as the water advanced steadily toward the crest, was Colonel Elias J. Unger, who had retired from business in Pittsburgh the year before and had only recently been named the club's president and over-all manager. He was living at the lake the year around now, in a modest farmhouse just beyond the spillway.

The Colonel had started life on another farm in Dauphin County, in the eastern part of the state. His father was a Pennsylvania German, as was his mother, who came from the big and well-known Eisenhower family. At twenty he got a job on the railroad and managed to work himself up from brakeman to conductor to superintendent of the Pennsylvania's hotels, from Pittsburgh to Jersey City, including the one at Cresson, where he was manager for a time and so got to know Carnegie and the others.

About the time the South Fork club was being organized, he had gone into the hotel business on his own in Pittsburgh and made even more of a name for himself. By 1888 he was well enough situated to buy the place on Lake Conemaugh and settle down to a quiet retirement in a glorious setting where there was also the added interest of a not very taxing job to keep him occupied, plus, in the summer months at least, the chance to keep up with his Pittsburgh friendships.

Unger had come a long way from Dauphin County. But even so, socially and financially, he was a noticeable cut below the other members of the South Fork fishing and hunting organization. His experience in hotel management, it would appear, had something to do with his position in the club.

The Colonel had returned home only the night before, after visiting friends in Harrisburg. When he got out of bed that morning at six, it looked to him, he later said, as though the whole valley below was under water, and he was baffled as to what it all meant. He put on his gum coat and boots and walked down the hill in front of the house, crossed the wooden bridge over the spillway and walked out onto the dam, where he began taking measurements of the rising water.

About eight thirty Unger's caretaker over at the club grounds, a man by the name of Boyer, came by in a spring wagon with D. W. C. Bidwell, who was on his way to South Fork intending to catch the 9:15 train to Pittsburgh. Bidwell, who evidently had had enough of the soaking weekend at the lake, stopped to ask Unger how things were going.

"Serious," answered Unger, who later that morning was heard to say that if the dam survived the day, he would see that major changes were made to insure that this sort of thing never happened again.

When Boyer got back from South Fork, which was sometime near ten, Unger sent him off to bring the Italian work crew down to the dam. He had decided to try digging another spillway at the western end, where he thought the hillside would be solid enough to keep the water from cutting through it too rapidly. There was brief disagreement over the idea, with some of the men protesting that the water would rip through any new wasteway so fast that the dam would quickly fail.

"It won't matter much," Unger said, "it will be ruined anyhow if I can't get rid of this water."

When it became clear that even the shallowest sort of ditch could barely be cut through the rocky hillside, Unger then set several of his men to work trying to clear away the debris which by now was clogging the iron fish screens in the main spillway and seriously reducing its capacity.

Among the bystanders taking all this in was a small fourteen-year-old boy with the big name of U. Ed Schwartzentruver, who, with some of his friends, had been there all morning in the rain watching the excitement. Seventy-six years later, sitting on his porch on Grant Street in South Fork, not quite ninety and nearly blind, he would talk about what he had seen that morning as though it had happened the day before.

"When this high water come down, there was all kinds of debris, stumps, pieces of logs, and underbrush and it started to jam up those screens under the bridge. The bridge was well constructed of heavy timber. There was a man named Bucannon up there, John Bucannon, who lived in South Fork. Well he kept telling Colonel Unger to tear out that bridge and pull out that big iron screen.

"But Colonel Unger wouldn't do it. And then when he said he

would do it, it was too late. The screens wouldn't budge, they were so jammed in by all that debris."

When John Parke came up onto the dam on horseback, he did what he could to exhort Unger's men to dig harder and faster, riding back and forth along the breast, shouting orders and moving men from one place to another when he thought it would do some good. But by eleven o'clock it was apparent to everyone that the lake was still advancing as fast as ever before. In fact, by eleven, the water was about level with the top of the dam and had already started to eat into what little had been thrown up by the plow and shovels. On the outer face, near the base of the dam, it looked as though several serious leaks had developed.

At this point, Colonel Unger decided that perhaps something ought to be done to warn the people in the valley below. The only way was to send a man down. There was a telephone line from the clubhouse to South Fork, but it was used only during the summer season and had not as yet been put in working order.

With all the rain there had been, the road to South Fork was in very bad shape, but John Parke made the ride in about ten minutes. Parke's relative youth, and the fact that he was not well known in the area, may account for the marginal success of his mission.

Furthermore, the first people to come from the dam to South Fork that morning, Boyer and Bidwell, had already told everyone that there was no danger of the water running over the top. So when Parke came splashing up Railroad Street with his warning, the news was both unexpected and perhaps seemed somewhat questionable. According to testimony made later by Bidwell, Parke stopped in front of George Stineman's supply store, which was across the street from the depot, and where a small crowd had gathered.

"I saw him come down there," Bidwell said, "and make a statement to the people standing about that the water was then running over the top of the dam, and there was very great danger of it giving way." Parke also told two men to go to the railroad's telegraph tower next to the depot and tell the operator to alert Johnstown. But soon after they left, Bidwell, according to one witness, began telling everyone that there was really nothing to get excited about.

The operator at South Fork that Friday morning was Miss Emma Ehrenfeld. She had come on duty at seven o'clock. It was about noon, she would later estimate, that a man came up into the tower "very much excited."

"Notify Johnstown right away about the dam," he said. "It's raising very fast and there's danger of the reservoir breaking."

"Who told you all this?" she asked.

"There's a man came from the lake," he said.

Emma was not quite sure how much to believe of his story. She had seen the man around town and thought his name was Wertzengreist, though she was uncertain about that too. But she said later, "He is a man that people generally don't have much confidence in, and for that reason, I scarcely knew what to do under the circumstances."

She was also hampered by the fact that her lines west were open only as far as the next tower, four miles down the river at Mineral Point. Beyond Mineral Point there seemed to be a break somewhere and so she had no direct contact with Johnstown.

The operator at Mineral Point, W. H. Pickerell, was an old hand along the Little Conemaugh, having been there at that same tower for some fifteen years. Emma decided to "talk" it over with him on the one good wire she had. She tapped out her problem and waited for an answer. Pickerell told her that the break to the west was caused by the poles falling into the river, and that though he had no way of getting Johnstown, he thought "it was a thing that there oughtn't to be any risks taken on." He said he would take the message and send it on to East Conemaugh by foot if someone should happen along the tracks below his tower.

So the two of them worked out a message addressed to the yardmaster at Conemaugh and to Robert Pitcairn in Pittsburgh.

"I wrote the message up," Pickerell declared weeks after, "and repeated it to her and asked her if that would do, and she said that was splendid—to send it that way. I doubled the message and waited and waited."

After a while a trackman came by. He had been sent from East Conemaugh to flag a landslide at Buttermilk Falls, to the west of Pickerell's tower. Pickerell gave him the folded message and sent him on his way back down the tracks. At Buttermilk Falls, the man, whose name was William Reichard, turned the message over to his boss, the foreman of the division, L. L. Rusher, who set out for East

Conemaugh after telling Reichard to go on back up to Mineral Point, in case there should be any more messages.

As it turned out, Rusher had only to go as far as what was known as "AO" tower, which was about a mile and half from Mineral Point and better than a mile upriver from East Conemaugh. From "AO" tower west the lines were still clear. Rusher gave the message to operator R. W. Shade, who sent it on immediately. And it was his message which was received by J. C. Walkinshaw at East Conemaugh and by agent Deckert in Johnstown sometime between noon and one o'clock.

In Pittsburgh, operator Charles Culp, at the Union depot later said he was the one who had received the message there and that he took it right over "and laid it on Mr. Pitcairn's table in front of him." Within an hour Robert Pitcairn, who had a special interest in the South Fork Fishing and Hunting Club, as well as the Pennsylvania Railroad, was sitting in his private railroad car on his way to Johnstown.

But the telegram drafted by operator Pickerell and Emma Ehrenfeld was only the first of three warnings sent down the valley by way of the Pennsylvania's telegraph system.

About twelve thirty, or sometime very shortly after John Parke reined up in front of Stineman's store, another rider was sent from South Fork to check the condition of the dam. His name was Dan Siebert. He worked for J. P. Wilson, who was superintendent of the Argyle Coal Company in South Fork and an old friend of Robert Pitcairn's. Wilson had been asked by Pitcairn some three years earlier to notify him at once if ever he saw any signs of danger at the dam.

Siebert borrowed Wilson's horse and was up and back from the dam inside of twenty-five minutes. He had stayed only long enough to see that near the center of the dam a glassy sheet of water, fifty to sixty feet wide, had started over the top. But Siebert did not seem especially concerned over what he had seen. He was, in fact, according to one witness, "perfectly cool about it."

Wilson, however, on hearing Siebert's report, turned to C. P. Dougherty, the Pennsylvania's ticket agent in South Fork, and asked him if he did not think that Mr. Pitcairn should be notified. When Dougherty hesitated, saying there was trouble with the wires downriver, Wilson took it to mean that Dougherty was reluctant to assume the responsibility of such a message and told him to sign his, Wilson's, name.

Whereupon Dougherty went over to the tower, taking along another operator, Elmer Paul, who was more experienced than Miss Ehrenfeld and who Dougherty thought might have better luck getting a circuit. Paul tried the wire for a few minutes but without success.

So again a message was sent as far as Mineral Point, where it was received at 1:52 by operator Pickerell, who gave it to William Reichard, who walked it down the tracks to "AO" tower. From there it was put on the wire to East Conemaugh, Johnstown, and Pittsburgh. The message read:

SOUTH FORK, MAY 31, 1889

R.P.   O.D. VIA MP & AO

THE WATER IS RUNNING OVER THE BREAST OF LAKE DAM, IN CENTER AND WEST SIDE AND IS BECOMING DANGEROUS.

C. P. DOUGHERTY

It was no more than thirty minutes later that J. P. Wilson came up to the tower himself to have Emma Ehrenfeld send still another warning. The rain was beating down terribly hard by then, and outside they could see the water of the Little Conemaugh and South Fork Creek raging across the flats just below the station. Over near Lake Street, South Fork Creek had flooded the first floors of several houses and the aspens along the banks were whipping about wildly in the wind.

Wilson had just heard that a young South Fork boy named John Baker had ridden down from the lake and said that the water had now cut a notch in the center of the dam. Without taking time to write anything down, Wilson dictated a message to Pitcairn which Emma Ehrenfeld put right on the wire.

SOUTH FORK, MAY 31, 1889

R.P.

OD

THE DAM IS BECOMING DANGEROUS AND MAY POSSIBLY GO.

J. P. WILSON

Wilson waited in the tower long enough to be sure the message had gotten at least as far as Mineral Point; then he warned Miss Ehrenfeld to be on the lookout up South Fork Creek and went out the door

The time was 2:25. By 2:33 the message had reached East Conemaugh. For some unknown reason, Pickerell this time had been able to get a circuit. Apparently, a wire that had fallen into the river lifted out somehow, and as Pickerell said, "All at once the wire came all right."

The message was through to East Conemaugh in a matter of minutes, and on to Johnstown and Pittsburgh. Agent Deckert in Johnstown would later state that he had received this particular message sometime near 2:45. He also concluded upon reading it that this time he had best telephone its contents across the way to Hettie Ogle, who ran the central telephone switchboard and Western Union office, just across the river at the corner of Washington and Walnut.

Mrs. Hettie Ogle was a Civil War widow who had been with Western Union for some twenty-eight years. At one o'clock the rising water had forced her to move, with her daughter, Minnie, to the second floor of the two-story frame building. Sometime near three she notified her Pittsburgh office of the condition of the dam as reported by Deckert and said that that would be her last message, meaning that the rising water was about to ground her wires. Then she put through a call to the *Tribune*, where editor Swank was still keeping up his running account of the day's events.

"At three o'clock," he wrote, "the town sat down with its hands in its pockets to make the best of a very dreary situation. All that had got out of reach of the flood that could, and there was nothing to do but wait; and what impatient waiting it was anyone who has ever been penned in by a flood and has watched the water rising, and night coming on, can imagine. . . ."

He described how the Stony Creek carried a live cow down from some point above Moxham and how she struck against a pier of the dislodged Poplar Street Bridge, where she managed to get a foothold for a while but finally, making a misstep, fell into the current and was carried off.

"At 3:15 the Central Telephone office called the *Tribune* up to say it had been informed by Agent Deckert, of the Pennsylvania

Railroad freight station, that the South Fork Reservoir was getting worse all the time and that the danger of its breaking was increasing momentarily. It is idle to speculate what would be the result if this tremendous body of water—three miles long, a mile wide in places, and sixty feet deep at the breast at its normal stage—should be thrown into the already submerged Valley of the Conemaugh."

But by 3:15 Lake Conemaugh was already on its way to Johnstown.

-3-

When John Parke had arrived back at the dam from his dash to South Fork, he was confident that his warning had been sent on down the valley. Along the way from South Fork he had passed two men struggling through the mud with a sewing machine, and one of them shouted to him, "We got the sewing machine out, if nothing else," which Parke took to be a very good sign. At the dam he found the water had already started sliding over the top, at the center, right above where the old culvert had been. It had taken no time for the water to wash across the little earth ridge that had been thrown up. Now, as he rode his horse out along the breast, the water crossing over the road there was a good six inches deep and getting stronger every minute. Within minutes the sheet of water was a hundred yards wide. But it was all concentrated at the center, clearly illustrating, Parke noted, that the dam dished a little.

It was now shortly after noon. At the western end the emergency spillway was running about twenty-five feet wide, but only slightly deeper than before. At the main spillway, where the water was roaring through six feet deep or more, the men had started to tear up the floor boards of the bridge and were attempting to remove a V-shaped floating drift made of logs with nails sticking out of them, which had been set out in the water twenty feet or so before the spillway to deter the fish even from venturing toward the screens.

The men were afraid to go out on the dam now, and so Parke rode across the breast alone, studying the effects of the overflowing water on the face of the dam. He saw that little gullies had already been cut between the riprap, but the damage was not as bad as he thought it would be.

For a brief moment he gave serious thought to the possibility of cutting another spillway through the dam proper, where there would be no problem digging deep and where the water would do quite a lot of the digging for them. It would have meant the end of the dam, of course; the water, he knew, would bore through any such cut, ripping the dam in two in no time. But by making such a cut near one of the ends, where the pressure was far less than at the center, the water, as fast as it might escape, would still go out a far sight slower than if the whole dam gave way all at once at the middle. It would have meant the certain destruction of the dam, but also far less damage below, he figured.

It would have been a terribly bold decision, and one which Parke alone would have been in no position to make. It would also be, he concluded, a foolhardy decision. Frightful damage would be caused for certain, and they would be responsible; furthermore, there would then be no way to prove that the dam had been about to break and that they had been left with no other choice. Indeed, Parke decided, there was no certainty at all that the dam was going to break. At which point he hurried to the clubhouse for his dinner.

When he returned to the dam, Parke found things had taken a decided turn for the worse.

Several big rocks on the outer face had washed away, and the water pouring across the top had cut a hole into the face about ten feet wide and four feet deep. As the water kept pounding down into this hole, he could see it slicing away at the face, a little more every minute, so that the hole began to take the form of a huge step.

There was absolutely nothing anyone could do now but watch and wait and hope. Parke, Unger, all of them, just stood there looking at the water and the valley stretching away below.

The rain-drenched crowd gathered on both hillsides had grown considerably since early morning. The news of trouble had spread fast and wide.

Some men had more or less smelled trouble hours before. George Gramling, who had a sawmill on Sandy Run, started off for the dam about eight in the morning, along with Jacob Baumgardner and Sam Helman. The Gramling mill was operated by a small dam which had broken about seven. If a small dam washed out that early, the men reasoned, what might a big dam do later on?

But mostly people were just curious. The Reverend G. W. Brown, pastor of the South Fork United Brethren Church, for example, like nearly everyone else in the neighborhood, had heard rumors of trouble all day and decided finally to go up to see for himself. When he arrived at the dam it was about ten minutes to three. There was no one actually out on the dam then, just at the ends, and the water was pouring over the breast.

Minutes later he saw the first break. He said it was "about large enough to admit the passage of a train of cars."

John Parke said that the break came after the huge "step" had been gouged back into the face so near to the water that the pressure caved in the wall.

Ed Schwartzentruver called the first break "a big notch."

"It run over a short spell," he said, "and then about half of the roadway just fell down over the dam.

"And then it just cut through like a knife."

Colonel Unger said the water worked its way down "little by little, until it got a headway, and when it got cut through, it just went like a flash."

Unger's man Boyer said, "It run over the top until it cut a channel, and then it ran out as fast as it could get out. It went out very fast, but it didn't burst out."

John Parke said, "It is an erroneous opinion that the dam burst. It simply moved away."

According to Ed Schwartzentruver, "The whole dam seemed to push out all at once. No, not a break, just one big push."

The time was ten after three.

IV

Rush of the torrent

-*1*-

There were men on the hillsides near the dam who had seen what the force of water could accomplish in mining operations, how a narrow sluice could scour and dig with the strength of a hundred men. Actually anyone who had lived in the area long enough to have seen even the spindliest of the local creeks in April had a fair idea of hydraulics at work. But no one who was on hand that afternoon was prepared for what happened when Lake Cone- maugh started for South Fork.

"Oh, it seemed to me as if all the destructive elements of the Creator had been turned loose at once in that awful current of water," Colonel Unger said.

When the dam let go, the lake seemed to leap into the valley like a living thing, "roaring like a mighty battle," one eyewitness would say. The water struck the valley treetop high and rushed out through the breach in the dam so fast that, as John Parke noted, "there was a depression of at least ten feet in the surface of the water flowing out, on a line with the inner face of the breast and

101

sloping back to the level of the lake about 150 feet from the breast."

Parke estimated that it took forty-five minutes for the entire lake to empty, but others said it took less, more in the neighborhood of thirty-six or thirty-seven minutes. In any case, later studies by civil engineers indicated that the water charged into the valley at a velocity and depth comparable to that of the Niagara River as it reaches Niagara Falls. Or to put it another way, the bursting of the South Fork dam was about like turning Niagara Falls into the valley for thirty-six minutes.

A short distance below the dam stood a farmhouse belonging to George Fisher. Fisher, who had been warned that the dam was about to go and had managed to escape from the house with his family only minutes before, saw everything he owned vanish in an instant.

Huge trees were snapped off or uprooted one after another and went plunging off in the torrent. When the flood had passed and the hollow was still again, the hill opposite the dam had been scraped bare for fifty feet up. Every bush, vine, every tree, every blade of grass, had been torn out. All that remained was bare rock and mud.

The water advanced like a tremendous wall. Giant chunks of the dam, fence posts, logs, boulders, whole trees, and the wreckage of the Fisher place were swept before it, driven along like an ugly grinder that kept building higher and higher.

At Lamb's Bridge, the little bridge itself as well as George Lamb's home were destroyed as swiftly as everything else. Lamb had been afraid of the dam but had not fled to higher ground until he heard the roar of the flood bearing down on him. He made a frantic effort to save two pigs but gave it up and got to the hillside with his family in time to see his house climb the face of the water, which, because of the narrowness of the valley at that point, was about sixty feet high. He watched the house roll and toss momentarily; then it was flung against the near hill and smashed to splinters.

From where they were the men at the dam could see all this happening as the water raged through the immense gash below them. But just beyond Lamb's Bridge the valley turns sharply to the right and disappears. So now they could only stand there, the rain beating down, and imagine, as much as that was possible, the things taking place beyond that turn.

The road to South Fork had disappeared, and with most of the dam gone, there was no way back to the clubhouse except the long way, clear around the lake, through flooded woods and fields where the mud would be impossible. So they stayed on, watching the level of the lake sink rapidly down and down, until there was nothing to see but hundreds of acres of dark ooze cut through by a violent yellow stream.

Colonel Unger lasted only a short time after the dam failed. He collapsed and had to be carried to his house and put to bed. His work crew, which had been hanging back nearby, waiting for his next orders, then climbed down to where the lake had been and with blankets and baskets and cold bare hands began scooping up the fish that were flopping about in the muck.

-2-

Emma Ehrenfeld was sitting with her back to the window in the telegraph tower just down from the South Fork depot. She was talking to H. M. Bennett, engineer of the 1165 freight from Derry, and S. W. Keltz, the conductor. The men had left Derry, halfway to Pittsburgh on the main line, the evening before and had been up all night, delayed first at East Conemaugh until five that morning, then held at South Fork since eight.

Tired, cold, rain-soaked trainmen had been coming in and out of the tower most of the day, climbing the stairs to ask about news from up the line or just warming themselves by the coal stove on the first floor.

Miss Ehrenfeld had held the *Chicago Limited* west of the bridge, on the other side of the Little Conemaugh, according to the orders she had received that morning; but with all the talk going around about trouble at the dam, the engineer had grown uneasy about his train standing over there, right where the flood might come. There had been a number of opinions on what to do, and then, after noon, the engineer got up and said he was going to bring the *Limited* across, orders or no orders.

After that there had been more speculating about the bridge. The conductor on the *Limited* had noticed cracks in one of the piers. The division foreman had been sent for, and when he came down from his house up the tracks and said the piers had looked that way for some time, they cut the *Limited*'s helper loose and ran it across first, very slowly, just to be sure. The *Limited* followed after and pulled up past the tower and the depot, a half mile or so. By the time that was done with, it was shortly before three.

Very soon after, Emma Ehrenfeld went downstairs to see about the stove. The men had been firing it up so that her little room upstairs was growing uncomfortably warm. The *Limited*'s engineer had come in again and was sitting there having a smoke, trying to dry off some. She passed the time with him for a few minutes, banged shut the door, and went back up the stairs to her desk.

From where he was sitting beside her, H. M. Bennett could see the northeastern corner of town neatly framed by the rain-spattered window. In the immediate foreground were the Pennsylvania tracks; just beyond them was Railroad Street, with Stineman's store to the left and the big Stineman house and Pringle's drugstore on either side of it. To the right was the turn where Railroad became Lake Street and headed uptown and out of sight, toward the road to the dam. And way over to the right, on the other side of the coal tipple and the planing mill, he could see South Fork Creek, flooding across the lowlands, through the trees, and reaching among the houses nearest its banks.

Suddenly Bennett noticed distant figures racing toward the hill. He jumped up and rushed to the window.

"Look at the people running!" he said. "I wonder what's wrong?"

The other two went immediately to the window and noticed that several people going by in the street below seemed to be shouting something.

As Miss Ehrenfeld later recalled, Bennett said something about the reservoir and that they ought to get out. Then they saw it coming, spread across the full width of the valley.

Situated as they were, only a few hundred feet from where the creek emptied into the Little Conemaugh, they were in about as good a place as any to see up the valley; but even so, they could not see very far because the abrupt hillside to which the town clung blocked off most of their view. When the water came into sight, it looked very close and enormous.

"It just seemed like a mountain coming," Emma Ehrenfeld said.

Conductor Keltz described it as more like a large hill rolling over and over. He judged it was about a hundred feet high.

The two men turned and dashed down out of the tower. Miss Ehrenfeld was right behind them ("without waiting to get my hat or anything"). She raced down the tracks, crossed over to the stairs that led to the coal tipple, ran to the top, and from there followed the crowd running toward the back alleys that led to higher ground.

Bennett and Keltz had started for the hill with her, but remembering the fireman and brakeman, who were asleep in the engine of their train on a siding on the other side of the river, they turned and ran for the bridge.

They made it to the engine, cut it loose, and with the little steam they had, came rolling out of the siding and back across the bridge, heading directly toward the oncoming flood with what looked like no better than an even chance of making safe ground only a few hundred yards away where the tracks swung hard to the left past the station.

Contrary to Keltz's estimate, the wall of water closing down on them was probably no more than forty feet high. It was moving straight for the bridge at a rate of perhaps ten to fifteen miles an hour and was driving before it a mass of debris that now included acres of trees, two or three small bridges, numerous mangled houses, dead animals, and rubbish beyond description.

About 200 yards from the bridge the water claimed its first human life. Michael Mann, an English coal miner and self-styled preacher who was known in South Fork as "The Reverend," had ignored every warning to leave the shanty he lived in on the banks of South Fork Creek. His body was found a week later, a mile and a half downstream. It was half-buried in mud, stripped of all clothing, and so badly decomposed that it could not be moved. As a result the last remains of the man who would be remembered in the valley as "The First Victim" were put into a hole nearby, covered over, and left unmarked.

The flood crushed right through the planing mill, wrenched the bridge from its piers, bent it as though it had been built with an elbow in the center, and then plowed head on into the mountain on the north side of the Little Conemaugh.

Engineer Bennett's locomotive meanwhile had escaped just about untouched. It had gotten almost to the station when another escaping freight pulled out of a coal siding and blocked the way. The next thing Bennett knew, a huge tree, evidently an advance fragment of the debris, smashed into his locomotive and pitched it halfway off the track. With the water almost on top of them, Bennett, Keltz, and the two others (they were both very much awake by now) jumped to the ground and scrambled onto the other train just as it started pulling away. Seconds later the full brunt of the flood roared past behind them.

But when it was all over remarkably little damage had been done in South Fork. Stacked on the hillside as it was, the town was almost entirely out of reach of the onslaught. Along with the bridge and the planing mill, some twenty other buildings and houses were destroyed. The bridge, which had been thirty-five feet above the normal water level, was dumped 200 yards *up* the Little Conemaugh, carried there by the violent backwash created when the water hit the mountainside. There were a few pieces of machinery to be found where the planing mill had been, but that was about all. There was a stone foundation marking where one store had been. A grocery and barbershop went sailing off. J. P. Wilson's stable containing two mules, a horse, and a cow landed behind the depot with the animals unhurt.

Station agent Dougherty's house was tossed into a gully, a total wreck. The depot itself had bobbed up several feet and swirled out over the tracks a ways. Then when the water rushed off downstream, it drifted back again and settled down almost precisely where it had been before, secured by a tangle of telegraph wires and only a little out of plumb.

The coal tipple was destroyed and so was the telegraph tower.

And that was about the size of it, except that there had been three other deaths.

A young man named Howard Shafer had been helping clear the jam of rubbish that had collected under a small bridge on South Fork Creek. When the water came he was unable to climb the steep bank fast enough.

The other two lost were Thomas Kehoe and Thomas Henderson, another fireman and a brakeman on Bennett's 1165 Derry freight. They had been asleep in the caboose when Bennett and Keltz had cut the engine loose to make their dash over the bridge. The caboose along with four other freight cars was carried away.

Past South Fork the water raged along the valley of the Little Conemaugh, between sharp, wooded bluffs that sent the riverbed swerving back and forth on its way west. A straight line from South Fork to Johnstown would be nine miles, but by the river route the distance was about four miles farther.

For the first mile or so beyond South Fork the valley runs on a comparatively even line and is nearly 1,000 feet deep. There were no houses, only the railroad, which skirted along the northern banks of the river about forty feet above the normal water level.

The flood ripped the railroad to shreds, tore out ties, twisted steel rails into incredible shapes, and swallowed up whatever equipment happened to be standing along the way.

A mile down from South Fork the valley narrows abruptly. There the rough hillsides squeezed the great mass of water so that its front wall grew to perhaps seventy to seventy-five feet high. Then, a half mile farther still, the river turns sharply south, traveling nearly two miles out of its way to form an oxbow which is only a matter of yards across. It was here, at the end of the oxbow, that the water smashed into its first major obstacle, a tremendous stone viaduct which had been built more than fifty years earlier to take the old Portage Railroad across the Little Conemaugh and which was still used for the main line of the Pennsylvania.

The viaduct was one of the landmarks of the country. It stood seventy-five feet high and bridged the river gap with one single eighty-foot arch. Even the biggest locomotives looked tiny by contrast as they chugged across it on their way up the mountain. Faced with a tawny-colored local sandstone, it was, as one engineer said, "a substantial and imposing piece of masonry," which had been built by "an honest Scotch stonemason" named John Durno from a design worked out by the same Sylvester Welsh who had picked the site for the reservoir.

The bridge had been built to save running the railroad clear around the oxbow. A cut had been made across the oxbow, a distance of less than a hundred feet, and the tracks had been run through it to the bridge. At the eastern end of the cut, where the river bends off to the south, the tracks were about twenty feet above the normal water level. But at the western end, where the tracks started over the bridge, they were seventy feet above the river. Thus the river's big two-mile loop to the south accomplished a drop of some fifty feet in elevation, which could have been achieved in less than a hundred feet, if the water were to take the path of the railroad cut.

When the flood hit this dividing point, part of the giant wave rushed through the cut at a depth of about twenty feet and plunged down over the top of the viaduct and into the deep river gully below, sweeping with it tons of debris which piled on top of the bridge or wedged between its huge arch.

Meanwhile, the rest of the water, and by far the greater proportion of it, crashed along the longer and more tortuous course of the river bed, heaped up to a height of perhaps seventy feet by the narrow channel and gathering before it the shredded refuse of two miles' worth of heavy timber, rock, and mud. Perhaps six or seven minutes passed before it swung around the last big bend before the bridge. When it struck, it was almost as high as the bridge itself.

The bridge held momentarily. There was an awful booming crunch as debris piled against stone and virtually sealed off the already clogged arch, and the water surged back and forth, seething with yellow foam, mounting up and up until it was nearly eighty feet high. And then it started spurting over the top of the bridge, gushing between the boulders and mangled railroad cars, the broken planks, ties, and tree stumps that had been dumped there.

Now, for a brief instant (no one knows exactly how long it lasted), Lake Conemaugh formed again some five and a half miles downstream from its original resting place. It gathered itself together, held now by another dam, which however temporary was nonetheless as high as the first one; and when this second dam let go, it did so even more suddenly and with greater violence than the first one. The bridge collapsed all at once, and the water exploded into the valley with its maximum power now concentrated again by the momentary delay.

A mile or so beyond the bridge was the white frame village of Mineral Point, consisting of some thirty houses set in a row along a single street, Front Street, which ran parallel with the river on the north side of the river. It was a pretty little place, quiet, clean, tucked at the foot of the mountainside.

The river there was quite shallow and filled with rocks. The water was quick and bright, and its steady rushing among the rocks was the first sound people heard when they woke up in the morning and the last they heard as they dropped off to sleep at night. Except for the railroad, which ran along the opposite side of the river well above the roof line of the houses, Mineral Point looked as though it might have been a thousand miles from civilization. The air smelled of the river and of sweet, fresh-cut timber at the sawmill and furniture factory, the town's sole supporting industry, which stood at the far end of Front Street.

The people who lived in Mineral Point had names like Reighard, Page, Sensebaugh, Gromley, Byers, and Burkhart, and there were perhaps 200 of them, if you counted some of the outlying families that picked up their mail there. The houses all faced the river and had deep lots, running back to where the woods started at the base of the hill. Fruit trees and truck gardens grew wonderfully in the moist soil put down by the river over long geologic ages.

Nothing much out of the ordinary had ever happened in Mineral Point. There had been a murder there once. A woman who was new to town and lived off to herself was killed by another stranger, a miner from over in the hard-coal country named Mickey Moore. The accepted story was that he was one of the Molly Maguires and that was the reason behind the killing. He had to carry out some dark oath. But that had been several years back. Mickey Moore had disappeared and life had gone on about as it always had, except that no child liked to stay out in the woods very long after dark. "Mickey Moore will get you," they said to each other, "sure as anything."

Beyond the last house, past the sawmill and around another bend or two in the river, and up on the opposite bank beside the railroad, was Mineral Point tower.

"I was sitting in the tower, and all at once, I heard a roar," W. H. Pickerell testified later. "I looked up the track, and I seen the trees and water coming. I jumped up and throwed the window up, and climbed out on a tin roof around our office and walked around on it, and I saw the driftwood coming around the curve, and the channel filling up and running over the bank, and I heard voices; I could hear somebody hollowing, but I couldn't see them, and I walked around until the drift came down, and looked out, and perhaps one third of the distance in the river, I saw a man standing on a house roof. He looked over and seen me and recognized me.

"He says, 'Mineral Point is all swept away, and the people swept away, and my whole family is gone.' I says, 'Is that so?' and I says, 'Do you know anything of my family?', and he says, 'No, I don't; I think they were all drowned.'

"Christ Montgomery was his name, and I says, 'Cheer up, Christ, don't give up; as long as you're on top, there's hope!'

"I didn't more than have the words out of my mouth until the drift he was riding made a straight shoot for the shore, and struck one hundred or one hundred and fifty yards west of my office where the river made a short turn, and went all to pieces; shingles flew right up in the air.

"He got out all right. He grabbed into the bushes just about the time it struck and I didn't see anything of him for a breath, and then he crawled up out of the bushes. After I cheered him up, and told him not to give up, that there was hope for him as long as he was on top; I turned around to walk into my office on this tin roof. I didn't have more than fifteen feet to walk, but I almost fainted when he told me my whole family was drowned. I turned right around to come in the office, and as I climbed toward the window, I looked and saw the house roof striking shore and seen him light, and saw him crawl up on his hands and knees, and saw he was saved, and when I looked above, there was a regular mountain of water coming. He was probably ahead of the main body of water a little.

"I started without coat or hat, and as it was pouring down

raining at the time, I turned around to get my coat and hat, and ran with them in my hand onto the opposite side of the track onto a high bank, and when I looked up the track, the wave wasn't more than a hundred yards off, and I beckoned for this man to get off the track. He wasn't looking for it to come down the track, and he got out on the track ahead of it, and came pretty near getting caught the second time."

Pickerell did not get caught at all nor, as things turned out, did any of his family, which was true of almost everyone else in Mineral Point.

The water had been coming up so fast that morning and during the early afternoon that most families had long since pulled out to higher ground by the time the flood fell on Front Street. First floors had been part way under water from about noon on, and there was no seeing the street or the riverbank. Picket fences, chicken coops, and backhouses had been drowned or had floated away as early as dinnertime.

But when the flood came, the wall of water swept through in such a way that it left almost nothing to suggest that there had ever been such a place as Mineral Point. The town was simply shaved off, right down to the bare rock.

The number of deaths came to sixteen, and quite a few people, like Christ Montgomery, went racing off on wild downstream rides astride their own rooftops. Christopher Gromley and his son traveled four miles before they were able to leap safely to shore; and three hours later, when they finally made it back to where Mineral Point had been, they found that all the other members of their family, Mrs. Gromley and six more children, were dead.

The water moved straight on down the valley, picking up a little speed wherever there were fewer turns to eat up its momentum and slowing down wherever the course began twisting again.

Estimates are that, in some places, it may have been moving as much as forty miles an hour. Theoretically, if its weight and the average decline in elevation (thirty-three feet per mile) are taken into account, it had a speed of nearly ninety miles per hour. But the friction created by the rough terrain and the rubbish it pushed before it cut that speed drastically. What is more, its over-all rate of advance was extremely fitful. The wall of debris and water came on not steadily but in an irregular series of thunderous checks and rushes.

At times, eyewitnesses said later, the debris would even clog the path enough to bring the whole thing to a momentary standstill. All the crushed and tangled sweepings from the dam down would lock clear across the valley, seeming almost more than the millions of tons of pressure from behind could budge.

But then the whole seething mass would burst apart, with trees and telegraph poles flying into the air, as though blasted by dynamite, and the water would rush forward again, even faster. And as it moved on, the water kept on tossing logs and roots above its surface, as though the whole mass were full of life.

The friction set up by the terrain and the debris also caused the bottom of the mass of water to move much slower than the top. As a result the top was continually sliding over the bottom and down the front of the advancing wall, like a cake of ice across a slick board. The water, in other words, was rolling over itself all the time it was pressing forward, and this caused a violent *downward* smashing, like a monstrous surf falling on a beach, that could crush almost anything in its path. A man caught under it had no chance at all. In fact, one of the major problems later on would be finding the bodies that had been pounded deep down into the mud.

Work train Number Two out of East Conemaugh was standing on the track nearest the hillside about a half mile upstream from the Conemaugh yards, at a place called Buttermilk Falls. The engineer sitting inside the rain-soaked cab was a friendly looking man with a round face and a dashing set of muttonchop whiskers. His name was John Hess.

Normally he never worked east of the yards. His division ran west from Conemaugh, as far as Sang Hollow, which was three miles below Johnstown. But today, with trouble almost everywhere along the line, help was being sent wherever it was needed.

Hess had gone to his regular engine as usual that morning and had been told to take a work crew down to Cambria City to clear a slide. His conductor was R. C. Liggett, his fireman, J. B. Plummer.

They had gotten through to Cambria City without any problems and worked there until nearly eleven, when an order came through to go clear up the valley to a landslide at Wilmore, on the far side of South Fork. At Johnstown and East Conemaugh there had been delays of twenty minutes and more, but sometime between noon and one they had started out of the yards, running east along the Little Conemaugh on the track farthest back from the water against the hillside. Less than a mile out they passed a place where a good hundred feet of track on the right had fallen off into the river. Beyond "AO" tower they came up on a flagman.

"I stopped to let him on," Hess recalled later, "and he says, 'You can't go any further.' And I asked him why, and he says, 'The north track is in the river and I don't believe the one you're on is safe,' and I says, 'Whereabouts?' and he says, 'Right through the big cut.' We went through the big cut to where the washout was, and seen it was badly washed, and I says to the conductor, 'I guess we'll have to take it afoot from here, and see where it is safe.' The conductor is an old experienced man, and he looked at the track we were on, and he says, 'It isn't safe, I won't run over that.' It was washed up to the ends of the ties and underneath the track, and undermined it; the ballast was still sticking to the ties; the ties seemed to be holding it up. He says, 'That isn't safe at all,' and we walked on up to Mineral Point, the next tower, and were going to report there but the operator told us he had no communication except with South Fork."

The operator, W. H. Pickerell, also told them about the messages which had been coming through from South Fork concerning the dam.

The men tramped back down the tracks to "AO" tower, where they took time out to eat. When they finished, it was about two o'clock and there was another message from East Conemaugh about a slide at Buttermilk Falls.

"We came down there," Hess said, "and found the track that we had went up on. The conductor thought at first it was unsafe, and we walked down over it and left the engine above it, and he suggested to cut a couple cars off—we had seven empty flats and the cabin ahead of our engine, and he suggested to cut off a couple cars and run them over to see whether it was safe, and probably we could bring the rest over. So we sent a man with two cars down over this dangerous place, and the bank didn't appear to slip much, and I brought the engine and rest of the train over. That left us on

the Conemaugh side of this washout. I went down and the brakeman coupled up those cars that they had sent down ahead, and the
conductor took the men with their shovels and went back to the
slide about one hundred yards back of where we were laying.

"I don't suppose we had laid there more than twenty minutes
until we heard the flood coming. We didn't see it but we heard the
noise of it coming. It was like a hurricane through wooded country."

Conductor Liggett heard the sound and thought he saw the
tops of the trees bend on the flat upstream between the railroad and
the river.

"And I says to the men, 'We'll get away from here,' and I still
looked, and then I was satisfied there was something coming. I
couldn't see any rubbish or drift, but I saw there was a commotion
among the green timber."

He shouted at the men to run. They dropped their tools and
started down the tracks looking for a place where they could climb
out of the way. But the rocks were too steep. They had to keep
running, 200, 300, nearly 400 yards before they found a path.

Hess and Plummer still could see nothing, but according to
Plummer, Hess said, "The lake's broke," and with that he put on
steam, tied down the whistle, and with their gravel cars clattering
along in front, they went shrieking toward East Conemaugh and
the railroad yards where the two sections of the *Day Express* stood
waiting.

The Hess ride into Conemaugh would be talked about and described in books and magazine articles for years to come, with Hess
in his engine (Number 1124), blazing down the valley, the water
practically on top of him, in an incredibly heroic dash to sound the
alarm.

Hess himself said afterward, "I didn't know what else to do. I
didn't see what else I could do."

He also said that he never did see any water, never waited
around that long. Moreover, Plummer estimated that their top
speed as they rounded the bend into the yards was no more than
twelve miles an hour, which, he said, was the best they could do
considering the load they were pushing, the condition of the tracks,
and the fact that they had no idea which way the waiting trains on

the other side of the blind turn might have been rearranged in their absence.

It was Hess's intention to keep right on going through the yards, clear to Johnstown, if the track was clear. But it was not. Plummer's guess was that no more than two minutes passed after they had pulled to a stop until the flood came.

"My brother was up on the bank and saw it coming," Plummer said. "I didn't see it coming at all; he saw it coming though and saw where it was, and he ran down and grabbed hold of me and gouged Hess with his umbrella, and told us to run."

With their whistle still screaming the two men jumped from the cab and started for the hillside.

A locomotive whistle was a matter of some personal importance to a railroad engineer. It was tuned and worked (even "played") according to his own particular choosing. The whistle was part of the make-up of the man; he was known for it as much as he was known for the engine he drove. And aside from its utilitarian functions, it could also be an instrument of no little amusement. Many an engineer could get a simple tune out of his whistle, and for those less musical it could be used to aggravate a cranky preacher in the middle of his Sunday sermon or to signal hello through the night to a wife or lady friend. But there was no horseplay about tying down the cord. A locomotive whistle going without letup meant one thing on the railroad, and to everyone who lived near the railroad. It meant there was something very wrong.

The whistle of John Hess's engine had been going now for maybe five minutes at most. It was not on long, but it was the only warning anyone was to hear, and nearly everyone in East Conemaugh heard it and understood almost instantly what it meant.

–3–

For the passengers on board the eastbound sections of the *Day Express*, the delay in East Conemaugh had been a dreary, monotonous affair. It was going on five hours now since the two trains had pulled to a stop between the river and the little town.

The first few hours had not been entirely uninteresting. A

number of passengers had gone out to look things over. They went walking about in the rain, up and down the tracks, over to the depot or the telegraph tower to see if there was any word on how long they would be held there. Or they picked their way across the tracks to the riverbank where the crowds were gathered and several local men were making great sport of spearing things of interest out of the racing current. And on the other side of the tower, the township bridge looked as though it would go almost any time.

But when dinnertime had passed and there still seemed no end to the rain and any chance of moving on seemed even less likely, whatever spirit of adventure there had been faded rapidly. The passengers had nearly all returned to the trains. They passed the time as best they could in the dim afternoon light, with the sound of the pelting rain all around them.

Elizabeth Bryan of Philadelphia sat looking out the window, while beside her, her friend Jennie Paulson of Pittsburgh read a novel titled *Miss Lou*. The girls had been to a wedding in Pittsburgh the day before and were on their way to New York, each wearing a small corsage of roses. Another passenger, the Reverend T. H. Robinson, a professor at the Western Theological Seminary in Allegheny, was busy writing a diary of the day's events for his wife.

Others were doing what they could to amuse their children. Some slept. One elderly gentleman, feeling slightly ill, had had his berth made up and retired for the day. Still others gathered in small clusters along the aisles to talk about the storm and the rising river, service on the Pennsylvania, the dismal prospect of the night ahead, or the possibility of getting a decent meal somewhere.

There was talk too about the dam farther up the mountain that everyone had been hearing about. But there was not much concern about it.

"The possibility of the dam giving way had been often discussed by passengers in my presence," one man, a bank teller from New Jersey, was later quoted, "and everybody supposed that the utmost danger it would do when it broke, as everybody believed it sometime would, would be to swell a little higher the current that tore down through Conemaugh Valley. Such a possibility as the carrying away of a train of cars on the great Pennsylvania Railroad was never seriously entertained by anybody."

Another passenger said that though many people may have been uneasy and were keeping "a pretty good lookout for information," the porters comforted them "with the assurance that the Pennsylvania Railroad Company always took care of its patrons."

So far whoever was directing things in the yard had chosen to move them twice. Twice they had watched the river working in on the tracks where the two trains stood, twice they had been moved forward and toward the hill, to be farther away from the river, and both times they had seen the tracks fall off into the water very soon after.

Now they were on the last sidings next to town, as far from the river as it was possible to be. The second section was on the track beside the depot and closest to town. Then came the first section, on the next track toward the river; and on the other side of it, four tracks over, was the local mail train. The *Day Express* engines were standing about even with the depot, with Section Two a few cars farther forward. The last cars were nearly on line with the telegraph tower, which stood on the river side of the tracks.

In the caboose of the mail train, which was nearest of the three to the tower, a fire was going in the stove and the conductors and others of the train crews were sitting about keeping warm between turns at checking in at the tower.

Messages had been coming in and going out of the tower steadily since early morning, and included those from South Fork. One operator, D. M. Montgomery, was later quoted as saying that the South Fork warning was generally well known. "But of course," he added, "nobody paid any more attention to it than if there hadn't been one at all. I know I didn't for one. It seemed like a rumor and they didn't take any belief in it."

Charles Haak, another operator in the tower, and the one who had passed along the first message from South Fork to the yardmaster's office downstairs, said he did not pay much attention to the warnings either.

"I was a stranger there," Haak said. "I had only been there but eight months, and of course, I listened to other people around there, residents there, and there was talk about the dam breaking, and they said there had been rumors but it never came, and so I thought that was how it would be this time."

As for the decisions on which trains to put where, they were being made by J. C. Walkinshaw, the yardmaster, who had been on duty since six that morning.

Walkinshaw was forty-nine years old, a widower with five children. He had worked for the Pennsylvania since he was seventeen and had been in charge of the East Conemaugh yards for twenty-three years. In a book of short biographical sketches of long-time company employees published later by the railroad, Walkinshaw peers out of a small photograph with wide eyes framed by white hair and whiskers, looking rather astonished and not especially bright. Robert Pitcairn later said that though Walkinshaw suffered from consumption, and so was "not very efficient" as yardmaster and "not very able to stand the physical strain," he, Pitcairn, nonetheless considered him amply qualified to look after the company's interest.

With circumstances as they were, Walkinshaw was left with little choice on what to do with the trains. He could not send them to the east, up the valley, because of the washouts at Buttermilk Falls and farther on. Nor could he send them back down the line toward Johnstown, as there were now reports of washouts in that direction as well.

About all he could do was to keep moving them back from the river, which is what he did. But once he had them on the northernmost siding, he concluded that he had taken "every reasonable precaution" under the circumstances.

One other very possible choice, of course, was to move the passengers out of the trains to higher ground. But to ask that many people to go out into the cold wind and rain, into the muddy little town where there might well be problems finding enough shelter for everyone, seemed more than Walkinshaw was willing to do, even though he had full knowledge of the trouble at Lake Conemaugh and was heard by at least one witness to say that if the dam ever broke it would "sweep the valley."

Walkinshaw had been out several times, checking equipment, giving orders, looking at the river. From two o'clock to three there seemed to be no change in the water level. Apparently the worst was over. But then about 3:15 the bridge below the telegraph tower went, causing a great stir among the crowd. Hour by hour the current had eaten away at its foundations, until they let go and the whole thing just dropped down into the water. Sometime shortly after, Walkinshaw went into his office where his son handed him another message about the dam. Then, about a quarter to four, Walkinshaw decided to take a brief rest.

"I sat down and wasn't in the chair more than a minute until I heard a whistle blow," he recalled later.

"I jumped off my chair, and ran out and hollered for every person to go away off the road and get on high ground, and I started up the track.

"Just as I left the office, I saw the rear end of this work train backing around the curve. I started up toward the train, and the minute I saw the train stop, I saw the engineer jump off and run for the hill. Just that minute, I saw a large wave come around the hill."

Inside the trains there was considerable commotion when the whistle started blasting. People stood up and began asking what the trouble was. Two Negro porters came through, both looking very excited and when asked if this meant that the dam had broken, the first one said he did not know, and the second said he thought it had. Outside, a conductor ran along between the trains shouting, "Get to the hill! Get to the hill!"

The Reverend Robinson said that no one knew what was going on, but that he remembered telling a woman next to him that he thought there was no danger. Then he looked out of the window and saw the wave coming. It appeared to be about 300 yards away, but there was no water to be seen. As one man said, it looked more like a hill of rubbish than anything else. Some people said it looked to be fifty feet high and it was taking everything in front of it. Everyone started for the door.

On Section One, the train standing between the mail train and Section Two, nearly every passenger got through the doors as fast as possible, but several of them, seeing the mud and rain, turned back. Jennie Paulson and her friend Elizabeth Bryan decided to go back for their overshoes. An old minister from Kalamazoo, Michigan, and his wife saw the flood bearing down and returned to their seats inside. But most people jumped and ran.

"It was every man for himself and God for us all," a New Haven, Connecticut, man named Wilmot said later.

Once they had clambered out into the rain, the passengers from Section One were faced with an immediate problem. On the next track, directly between them and high ground, stood Section Two.

"I saw three ways before me," the Reverend Robinson wrote afterward, "climb over section No. 2 or crawl under it, or run down the track with the flood four car lengths and around the train. I instantly chose the latter. No one else followed me so far as I saw, but all attempted the other courses."

Robinson made it safely around the train, but between him and the town and the streets which climbed to high ground was a ditch running parallel with the last track. It was about ten feet wide and perhaps five feet deep and rushing with water the color of heavily creamed coffee. Fortunately for him, Robinson arrived at the ditch at a place where a big plank had been laid across it. He was over in seconds and on his way up a steep, mud-slick embankment toward the town.

But others hesitated at the ditch, or leaped, or fell in and floundered about desperately, panicked. A number of men stopped, then moved back several steps, got a start, and jumped across. George Graham, a doctor from Port Royal, Pennsylvania, made it over this way; then, feeling that he still had time to spare, turned back to see if he could help some of the others.

"Just to my left, into the ditch, armpit deep, I saw nine women and girls tumble. I instantly grabbed the hand of the first and quickly pulled her out; the meanwhile all the others reached for me at once. I succeeded in saving them all except one old lady." Wilmot, the New Haven man, also cleared the ditch, carrying his child in his arms. When he looked back to find out what had become of his wife, he saw her hesitating on the other side, while a man beside her shouted, "Jump, jump!" She jumped and made it, and they ran on.

Cyrus Schick, a prominent Reading businessman who, with his wife and her sister, had been on his way home from a long health tour in the west, fell headlong into the ditch, as did his sister-in-law, Eliza Stinson. Schick's wife saw him bob up out of the water but then lost sight of him in all the confusion. His body and that of Miss Stinson were not found for ten days.

On the other side of the ditch the streets were full of running, shouting people. One local girl, a pretty young schoolteacher named Kate Giffen, who lived with her family on Front Street, would later describe racing to her house to pick up a child and seeing the woman who lived next door standing out on her porch screaming. She was the wife of John Hess, and she was screaming that the locomotive whistle still blasting away in the yards below was her husband's.

The Reverend Robinson found himself all alone, pressing up a back alley.

"I ran to the second street, and, hoping I might be safe, I turned and looked. The houses were floating away behind me, and the flood was getting round above me. I ran on to the third street and turned again; the water was close behind me, houses were toppling over, and the torrent again pushing round as if to head me off."

He kept on running, and when he turned again, he was high enough to see most of the town and the river valley. He watched a railroad car break loose and bound off in the plunging water, with two men on top trying desperately to keep their balance, moving first to one side then to another, as they headed toward Johnstown. How many passengers there might have been inside he could not tell. Everywhere people were rushing this way and that, some ducking inside doorways, some going for higher ground, stumbling and falling in the muddy streets. As the wave hit Front Street, buildings began falling, one on top of another; some seemed to bounce and roll before they were swept downstream. Locomotives from the roundhouse went swirling about like logs in a millrace.

The big, brick roundhouse had some nine engines in it when the flood struck. There were also another nineteen or twenty engines elsewhere in the yards, machine shops, a lot of rolling stock, a coal shed, and the three passenger trains. When the wave struck, it was probably about twenty-eight to thirty feet high, though, understandably, it looked a great deal higher to anyone caught in its path. The roundhouse was crushed, as one onlooker said, "like a toy in the hands of a giant." The passenger trains were swamped in an instant. Section One was ripped apart and the baggage car and one coach were flung downstream and its Pullman coach caught fire. Yard engines went spinning off, one after another.

Section Two and the mail train both miraculously survived. Section Two had been standing on an embankment five to six feet high, which certainly had something to do with its good fortune. There were also some freight cars in front of it and a coal shed that toppled across the tracks and helped deflect some of the onrush. But it was the roundhouse which almost certainly did most of the deflecting, and the fact that the valley both curves sharply and broadens out at that particular point along the river undoubtedly contributed to the inconsistent behavior of the oncoming wave.

The destruction all around the trains was fearful. Forty houses along Front Street were taken away. The Eagle Hotel, the Central Hotel, the post office, the railroad station, several stores, at least half the town was destroyed. The only railroad track left was that under Section Two, the mail train, and a few other pieces of equipment that, for one quirk of fate or another, happened to survive.

Thirty locomotives, some weighing as much as eighty tons, were scattered anywhere from a hundred yards to a mile from where they had been standing. One locomotive boiler would be carried all the way to Johnstown. How many lives were lost was never determined exactly. But at least twenty-two passengers from the *Day Express* sections were killed, including Cyrus Shick and his sister-in-law, Jennie Paulson and Elizabeth Bryan, the minister's wife from Kalamazoo, and F. Phillips, one of the Negro porters.

In East Conemaugh and Franklin, which was the name of the cluster of houses across the river from the yards, the known death toll came to twenty-eight.

But when the flood had passed, the engine and tender and six cars of the second section were almost at the exact same place they had been since before noon. They had been shoved along the track some, maybe twenty yards downstream. There was debris jammed in around them; but the sixteen people inside, who out of fear or indecision or dumb luck had stayed on board, were as safe and sound as though virtually nothing had happened. The water had come up over the seats in several cars and the passengers were soaked to the skin and badly shaken by the experience, but the only fatalities from their cars were among those who had tried to make a run for it.

One such passenger was John Ross of New Jersey, who, it was said later, was about thirty-three years old and a cripple. He had been traveling in one of the sleepers of Section Two, a car in which no one had chosen to hang back. Ross struggled out with the rest and was having a terribly difficult time until one of the train crew, a brakeman named J. G. Miller, came running along, picked him up, and managed to carry Ross some fifty yards or more before he dropped him.

"I had to drop him," Miller said later, "to save myself. I saw it was either life or death with me, and I dropped him, and went for the hill."

The mail train, which had been standing on an even lower track and within no more than a hundred yards from the river, was also still intact, though it too had been shoved downstream quite a way. Like Section Two, the mail train had been partly sheltered by the roundhouse, but what seems to have saved it was the telegraph tower which fell right onto the engine just as it was being pushed past underneath, and pinned it down there until the water had passed. But unlike any of the other trains, there were no fatalities among its passengers. Everyone got off and onto the hill in time, thanks to the good sense of the crew and, perhaps in part, to the particular nature of the passengers themselves.

Like all the others milling about the yards that morning, the eighteen or so passengers on board the mail train had heard mixed reports about the dam. They were told that if it ever broke it would drown the valley, and they were told that it would raise the level of the river maybe a foot or two. They were told it would take the water one hour to get from the lake to East Conemaugh, and they were told that it would take three hours. But mostly they were told that the dam was an old chestnut and not to think any more of it.

But their conductor, Charles Warthen, decided to tell them everything he knew, which was not a whole lot more, but he at least made it sound serious. He also told them to get ready to move out at the slightest notice, which was something neither of the conductors of the other trains had chosen to do.

The trainmen had been sitting in the last car of the mail train, talking about the situation, but for some reason or other, S. E. Bell, who was conductor on Section One, and Levi Easton, conductor on Section Two, made no effort to warn their passengers. The likeliest explanation seems to be that they, like so many others, had no real fear of anything happening.

All but one or two of Warthen's passengers were from the *Night Off* company. When they were told what might be expected of them, they quietly went to work rounding up their belongings, and the women began pinning up their skirts.

About two o'clock C. J. McGuigan, brakeman on the mail train, had gone to the tower to ask if there was any further news, and the operator (which one it was he did not recall) said, "Nothing, only another message that the dam is in a very dangerous condition." Not knowing anything about the dam, McGuigan asked him what the consequences might be if the dam broke.

"He kind of smiled," McGuigan told the story later, "and said, 'It would cover this whole valley from hill to hill with water.' I got kind of frightened myself then, and I came right down, and told the passengers the second time to be on the lookout. . . . The ladies got frightened, and one of them wanted to know if they should not better go to the hills now, but the manager of the troupe said 'No, there is no danger yet' . . . The women seemed to be ready for it . . . I think they were very sensible people."

McGuigan then went back to the last car to the other crewmen. When the whistle began blowing, he ran to the passenger coach, shouting that the flood was coming, while conductors Bell and Easton took off for their trains, shouting the same thing.

"The women were sitting down, and the men were standing up, and they all had their grips and valises in their hands, and the men ran to the upper end of the car, and the ladies to the west end where I was.

"I assisted them out, and got up and looked through the train, and I couldn't see anybody on the train, and then I ran with two of the ladies, caught hold of their hands, and ran until we came to the ditch . . . and Miss Eberly, she refused to go into the ditch, and I threw her into it, and jumped down and assisted her up on the other side, and ran up the hill."

No one was lost, not even the baggagemaster, J. W. Grove, who decided to jump onto one of the yard engines standing about instead of trying for the hill. Every other loose engine in East Conemaugh was dumped over, driven into the hillside, or swept off with the flood, except the one he picked.

Brakeman McGuigan went about for some time after carrying a picture of Miss Eberly, who was the pretty, young star of the company and actually Mrs. Eberly. She in turn was quoted widely when she returned to New York and described the bravery of the trainmen.

Later the Pennsylvania Railroad, in an effort to establish exactly what had happened at East Conemaugh, conducted its own investigation, which would provide the one full account of the whereabouts of several dozen employees, the official decisions made before the water struck, and the personal decisions made when it was seen rounding the bend behind the Hess train. The study revealed several cases such as that of brakeman McGuigan, but it included many more where the reaction had been a good deal less coolheaded and quite a lot more human.

Samuel S. Miller, for example, was also a brakeman, on the first section of the *Day Express*, the one on which most of the fatalities occurred. Part of his testimony went as follows:

Q. Where were you when the big wave came?
A. I was partly up on the hill.
Q. What were you doing up there?
A. Well, I was told that it was coming, and I got up on the hill for my own safety. I had gone to the Agent at Conemaugh, he was in the office at Conemaugh station—
Q. Who is he?
A. E. R. Stewart—and I borrowed the key from him for the water closet at the station, and I went in the water closet, and I think I was reading a *Commercial-Gazette* at the time when I heard the big whistle, and not knowing of any freight moving, I first thought probably it might be a freight engine that was to assist first *Day Express* up the mountain; I thought maybe they were alarming the passengers to get on the train and wondered why it wasn't a passenger engine whistle. The next thought that came to me was that South Fork dam had broken. I made a hasty exit, and when I got outside, a young fellow came along and said that was what was wrong. He seemed to be in a great hurry, and I asked him if South Fork dam had broken, and he replied, "Yes, so people say," and it seems to me, I told him to run, and I ran too.
Q. You broke for the hill?
A. Yes, sir, I broke for the hill.

Q. You didn't go to your train?

A. No, sir; I got up on the hill probably 110 yards from the station, and looked back, and could see that the water had come. I could see that the water was between the houses at that time. I concluded I wasn't high enough, and I went up onto still higher ground.

Q. You didn't climb a tree?

A. No, sir.

Q. Why didn't you go to your train and help get your passengers out?

A. Well, for my own safety. From the descriptions I had heard, I concluded I had better be on the hill.

Q. You might have gone to your train if you had tried?

A. I could have, but the question was whether I could then have gone to the hill or not.

Q. You believed your life was in danger, did you?

A. Yes, sir.

Now several hundred freight cars, a dozen or more locomotives, passenger cars, nearly a hundred more houses, and quite a few human corpses were part of the tidal wave that surged on down the valley.

Before it had plowed through East Conemaugh, the water had cut along the valley below Mineral Point, crashing back and forth against the mountainsides as the river channel swung this way and that. A mile or so above East Conemaugh, at the place the railroad men called "the big cut," the Pennsylvania tracks again left the riverbank to take a short cut across another oxbow. Here again the flood had divided briefly, with part of it rushing headlong through the cut, while the rest went with the river on its two-mile loop off to the north. It was a course which sapped much of the wave's potential speed and energy. But from East Conemaugh on to Johnstown the valley opened up considerably and the river headed directly for its meeting with the Stony Creek. Past East Conemaugh the flood was on a straightaway, and there it began to gather speed.

Woodvale got it next. Woodvale was somewhat bigger than East Conemaugh, prosperous, new, and the pride of the Cambria Iron Company. It was a sort of model town, built by the company, and with its clean white houses it looked, as one man said, more like

a New England town. It was connected to Johnstown by a streetcar line that ran along its main thoroughfare, Maple Avenue, which
was far and away the prettiest street in the valley. Maple Avenue
was nearly a mile long and looked like a green tunnel that May.
The trees reached over the tracks where the little yellow streetcars
rattled by, their horses heading for the stable. When the flood had
passed, there would be no trace of Maple Avenue.

About 1,000 people lived in Woodvale. There was a woolen
mill, built by Cambria Iron, which employed several hundred
women. There was the Rosensteel tannery, two schoolhouses, some
churches, and no saloons (they evidently were contrary to the Iron
Company's idea of a model town).

Unlike East Conemaugh, Woodvale got no warning. It was all
over in about five minutes. The only building left standing was the
woolen mill, and there was only part of that. At the western end of
the town, the end almost touching Johnstown, stood the Gautier
works, part of it in Woodvale, part in Conemaugh borough. The
huge works sent up a terrific geyser of steam when the water hit its
boilers, and then the whole of it seemed simply to lift up and slide
off with the water. The tannery went and so did the streetcar shed,
along with eighty-nine horses and about thirty tons of hay. When
the water had passed, the town was nothing but a mud flat strewn
with bits of wreckage. There was only a tiny fringe of houses left
along the edges, on the foothills. There was not a tree, not a telegraph pole, not a sign of where the railroad had been. Two hundred and fifty-five houses had been taken off, and there was no way
of telling where they had been.

The official figure for Woodvale's dead would later be set at
314, which means that about one out of every three people in town
had been killed.

A number of people had tried to crawl under a freight train
that was blocking their way to the hill and had been crushed when
the water hit the train and it started moving. Dozens of others had
never made it out of their houses. At the woolen mill three men had
kept retreating to different rooms and higher floors as the big brick
building caved in piece by piece, all around them, until there was
only that small part which miraculously withstood everything that
was thrown against it.

When the wireworks broke up it contributed miles and miles

of barbed wire to the mountain of wreckage and water that, once past the wireworks, had only a few hundred yards to go until it struck Johnstown.

It was now not quite an hour since the dam had given way. The rain was still coming down, but not so hard as before, and the sky overhead was noticeably brighter.

In Johnstown the water in the streets seemed actually to be going down some. It had been a long, tiresome day in Johnstown, and the prospects for a night without gas or electricity were not especially cheerful, but by the looks of the water and the sky, the worst of it had passed.

Johnstown was still sparsely settled when this map was made about the time of the Civil War. Woodvale (top) and Kernville (lower right) were mostly vacant lots. By 1889 population in the area had tripled.

One of the few surviving photographs of the South Fork Fishing and Hunting Club shows two lake-side boathouses, boardwalks, a rowboat planted with flowers (at lower left) and several cottage fronts.

The Moorhead house, once among the finest "cottages" on Lake Conemaugh, stands today at the edge of St. Michael, a coal town that, years after the flood, grew up around the abandoned summer colony.
DAVID G. McCULLOUGH

The heroic "resident engineer" at the club, John G. Parke, Jr.

At left below, a small railing marks the crest of South Fork dam, the immense ends of which still stand above South Fork Creek. From railing to distant rooftops (the town of St. Michael) was once the northern end of Lake Conemaugh. Large roof is the old clubhouse, only half of which still stands. It is now a miners' bar and hotel.

DAVID G. MCCULLOUGH

UNIVERSITY OF PENNSYLVANIA

Andrew Carnegie

Henry Clay Frick

FOUR MEMBERS OF THE SOUTH FORK FISHING AND HUNTING CLUB

Philander C. Knox

Robert Pitcairn

Daniel J. Morrell   John Fulton

FOUR OF JOHNSTOWN'S LEADING CITIZENS

Tom L. Johnson   Captain Bill Jones

A wide-angle view of Johnstown taken on the day before the flood gives the valley a broader look than it actually has, makes the hills appear too low-lying, but shows such principal features as the Stony Creek (foreground), the Cambria works (where the smoke rises at far left), Prospect

This view was taken from the same point several months after the flood and shows the extent of damage to the lower part of Johnstown (at left) and Kernville. Not shown are Woodvale (which sits in the middle of the gap), where the whole town was wiped out, or Cambria City. Also, though

Hill (the rows of white houses to the right of the mills), and the gap (center) through which the wall of water rushed. Kernville is at right. Johnstown proper lies between the gap and the church steeples, and to the left of the river.

there are still hundreds of houses to be seen in the main part of town, only a relative few were still in one piece, or standing where they belonged. Green Hill, a refuge for thousands of flood survivors, rises on the right of the gap.

THE CAMERA SHOP, JOHNSTOWN

With dozens of displaced buildings and tons of debris piled up behind it, the big stone Methodist church (at upper left in the view above) stands unmoved, looking over the desolation of lower Johnstown (in the far distance) and the dim side of Prospect Hill (at upper right).

A favorite subject for the swarms of photographers who rushed to cover the disaster was the house belonging to John Schultz. It had been neatly skewered by a huge tree and then dumped down near the Point. Six people were in it when the wave hit. Miraculously they all came out alive.

The photograph at right is almost certainly a fake. Though hundreds of corpses were strewn among the wreckage, few were found looking quite so neat and clean as this barefoot "victim," and by the time the photographers arrived any body so well exposed would have been long since removed.

A view taken less than 24 hours after the flood shows the Cambria Iron offices (the big buildings on the left) and the swamped ruins of the city beyond. At far right is part of the depot. Two Cambria Iron locomotives stand at lower right, inside the high board fence that enclosed company grounds.

Weeks after the flood, workmen tackle the last remains of the debris piled against the Pennsylvania Railroad's massive stone bridge.

In a view from above the Stony Creek, looking toward the Point (the stone bridge and the mills can be seen faintly in the distance), a slim island of surviving houses stands amid acres of mud and ruin.

Gertrude Quinn, at the age of 5. This photograph was one of the few Quinn family possessions to survive the flood.

Victor Heiser as he looked at 19, three years after the flood, when he had left Johnstown to begin his college education.

For Richard Harding Davis, the flood was the first big assignment in a long, colorful career as roving reporter and author.

For Clara Barton, "Angel of the Battlefield," the flood was the first great test of her newly organized American Red Cross.

LIBRARY OF CONGRESS

With their houses swept away, their money and belongings all gone, many families "made do" in rough hillside shelters such as these, built and furnished with the best they could scavenge from the devastated valley.

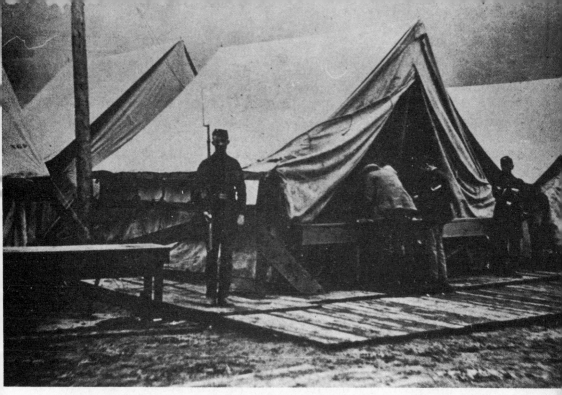

At the center of town, where the militia had set up camp, a young sentry stands guard while a survivor signs up for relief rations.

The body of a child is carried into the Adams Street schoolhouse, a temporary morgue where 301 bodies were recorded in the log books.

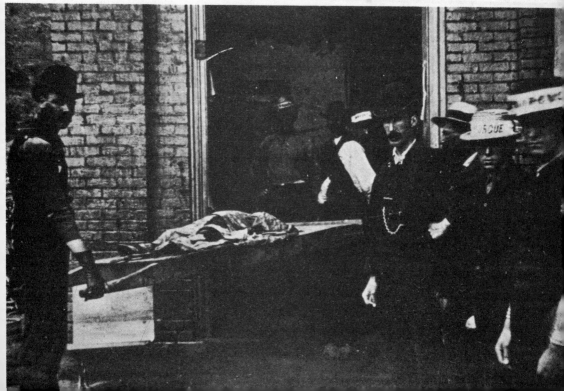

An artist's drawing of the broken dam, seen from inside the empty reservoir, shows the spillway (at far right) and the bridge that crossed it. The breach in the dam was about 420 feet across the top.

At the Pennsylvania depot survivors crowd one of the several commissaries set up by volunteers from Pittsburgh and other nearby towns.

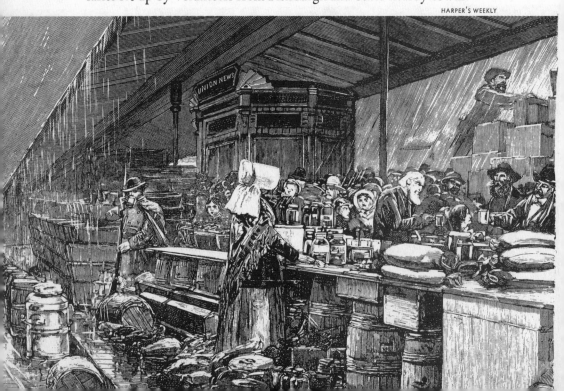

BROWN BROTHERS

Scenes such as the one above and at left below were repeated many times, as survivors searched among the dead, or suddenly found a lost loved one among the living. But the illustration at right below, supposedly depicting the demise of three "Huns" who had been caught robbing the dead, has no basis in fact. Popular as they were in the nation's press, lurid stories of rampant pillage and bloodshed were soon entirely discredited.

BROWN BROTHERS

DAVID G. MCCULLOUGH

At Grandview Cemetery, on a high hill above Johnstown, hundreds of plain marble headstones mark the graves of the flood's unknown dead.

V

"Run for your lives!"

=====

-1-

Most of the people in Johnstown never saw the water coming; they only heard it; and those who lived to tell about it would for years after try to describe the sound of the thing as it rushed on them.

It began as a deep, steady rumble, they would say; then it grew louder and louder until it had become an avalanche of sound, "a roar like thunder" was how they generally described it. But one man said he thought the sound was more like the rush of an oncoming train, while another said, "And the sound, I will never forget the sound of that. It sounded to me just like a lot of horses grinding oats."

Everyone heard shouting and screaming, the earsplitting crash of buildings going down, glass shattering, and the sides of houses ripping apart. Some people would later swear they heard factory whistles screeching frantically and church bells ringing. Who may have been yanking the bell cords was never discovered, but it was later reported that a freight engineer named Hugh Clifford had raced his train from above the depot across the stone bridge, his

145

whistle going the whole way; and a man named Charles Horner blew the whistle over at Harry Swank's machine shop.

Those who actually saw the wall of water would talk and write of how it "snapped off trees like pipestems" or "crushed houses like eggshells" or picked up locomotives (and all sorts of other immense objects) "like so much chaff." But what seemed to make the most lasting impression was the cloud of dark spray that hung over the front of the wave.

Tribune editor George Swank wrote, "The first appearance was like that of a great fire, the dust it raised." Another survivor described it as "a blur, an advance guard, as it were a mist, like dust that precedes a cavalry charge." One young man said he thought at first that there must have been a terrible explosion up the river, "for the water coming looked like a cloud of the blackest smoke I ever saw."

For everyone who saw it, there seemed something especially evil about this "awful mass of spray" that hovered over "the black wreck." It was talked of as "the death mist" and would be remembered always.

The fact was there had been something close to an explosion up the river, at the Gautier works, when the water rolled over the fires there, which undoubtedly accounted for a good part of what they saw. Horace Rose, who witnessed about as much as anyone, thought so.

At the first sound of trouble he had rushed to the third floor of his house on lower Main Street and from the front window could see nearly a mile up the valley. Only a few minutes before he had been playfully teasing his neighbors' child, Bessie Fronheiser, from another window downstairs, telling her to come on over for a visit. The distance between the two houses was only about five feet, so he had put some candy on the end of a broom and passed it over to her. That was so successful that he next passed across a tin cup of coffee to Bessie's mother in the same way. She was just raising the cup to her lips when the first crash came.

From the third floor Rose could see the long line of the rolling debris, stretching from hill to hill, slicing through the Gautier works, chopping it down and sending up a huge cloud of soot and steam.

The sight took his breath away. Once clear of the wireworks, the wave kept on coming straight toward him, heading for the very heart of the city. Stores, houses, trees, everything was going down in front of it, and the closer it came, the bigger it seemed to grow. Rose figured that he and his family had, at the most, two or three minutes before they would be crushed to death.

There would be slight differences of opinion later as to precisely when the wave crossed the line into Johnstown, but the generally accepted time is 4:07.

The height of the wall was at least thirty-six feet at the center, though eyewitness descriptions suggest that the mass was perhaps ten feet higher there than off to the sides where the water was spreading out as the valley expanded to a width of nearly half a mile.

It was also noted by dozens of people that the wave appeared to be preceded by a wind which blew down small buildings and set trees to slapping about in the split seconds before the water actually struck them. Several men later described how the wind had whipped against them as they scrambled up the hillsides, grabbing at brush to pull themselves out of the way at the very last instant.

Because of the speed it had been building as it plunged through Woodvale, the water struck Johnstown harder than anything it had encountered in its fourteen-mile course from the dam. And the part of the city which took the initial impact was the eastern end of Washington Street, which ran almost at right angles to the path of the oncoming wave.

The drowning and devastation of the city took just about ten minutes.

For most people they were the most desperate minutes of their lives, snatching at children and struggling through the water, trying to reach the high ground, running upstairs as houses began to quake and split apart, clinging to rafters, window ledges, anything, while the whole world around them seemed to spin faster and faster. But there were hundreds, on the hillsides, on the rooftops of houses out of the direct path, or in the windows of tall buildings downtown, who just stood stone-still and watched in dumb horror.

They saw the eastern end of Washington Street, the block where the Heiser dry-goods store stood, disappear in an instant. From there the wave seemed to divide into three main thrusts, one striking across the eastern end of town behind the Methodist

Church, one driving straight through the center, and the other sticking more or less to the channel of the Little Conemaugh along the northern side of town. Not that there was any clear parting of the wave, but rather that there seemed to be those three major paths of destruction.

East of the park, Jackson and Clinton streets became rivers of rubbish churning headlong for the Stony Creek. On Main and Locust, big brick buildings like the Hulbert House collapsed like cardboard while smaller wood-frame stores and apartment houses jumped from their foundations and went swirling away downstream, often to be smashed to bits against still other buildings, freight cars, or immense trees caught by the same roaring current.

Every tree in the park was torn up by the roots and snatched away as the water crossed through the center of town. John Fulton's house caved in, and other big places went down almost immediately after—the Horace Rose house, the John Dibert house, the Cyrus Elder house. The library, the telegraph office, the Opera House, the German Lutheran Church, the fire station, landmarks were vanishing so fast that no one could keep count of them. Then, perhaps no more than four minutes after the water had plunged across Washington Street, it broke past Vine on the far side of town and slammed into the hill which rises almost straight up to nearly 550 feet in back of the Stony Creek.

It was as though the water had hit an immense and immovable backboard, and the result was much as it had been at South Fork when the wave struck the mountainside there. An immediate and furious backwash occurred. One huge wave veered off to the south, charging *up* the Stony Creek, destroying miles of the densely populated valley, which, it would seem, had been well out of reach of any trouble from the valley of the Little Conemaugh. Other waves pounded back on Johnstown itself, this time, very often, to batter down buildings which had somehow withstood the first onslaught.

Houses and rooftops, dozens of them with thirty or forty people clinging on top, went spinning off on a second run with the current, some to end up drifting about for hours, but most to pile in to the stone bridge, where a good part of the water headed after striking the hill, and where eventually all the water had to go.

The bridge crossed the Conemaugh River downstream from

the Point where the Stony Creek and the Little Conemaugh come together. Past the bridge, another mile or so west, was the great Conemaugh Gap, the deepest river gorge between the Alleghenies and the Rockies and the flood's only way out of the mountains. But the bridge was never hit by the full force of the water. It had been built far enough down from the Point so that when the wave went grinding over Johnstown, it was shielded by Prospect Hill, and after the wave broke apart against the mountainside, the bridge had to withstand the impact of only a part of the wave.

As a result the bridge held. Had it been in the direct path and been struck full force, it would have been taken out just like everything else. But as it was, the mountainside took the brunt of the blow, the bridge survived, and the course of events for the next several hours went very differently.

Debris began building rapidly among the massive stone arches. And now it was no longer the relatively small sort of rubbish that had been clogging the bridge most of the day. Now boxcars, factory roofs, trees, telegraph poles, hideous masses of barbed wire, hundreds of houses, many squashed beyond recognition, others still astonishingly intact, dead horses and cows, and hundreds of human beings, dead and alive, were driven against the bridge until a small mountain had formed, higher than the bridge itself and nearly watertight. So once again, for the second time within an hour, Lake Conemaugh gathered in a new setting. Now it was spread all across Johnstown and well beyond.

But this time the new "dam" would hold quite a little longer than the viaduct had and would cause still another kind of murderous nightmare. For when darkness fell, the debris at the bridge caught fire.

No one knows for sure what caused the fire. The explanation most often given at the time was that oil from a derailed tank car had soaked down through the mass, and that it was set off by coal stoves dumped over inside the kitchens of mangled houses caught in the jam. But there could have been a number of other causes, and in any case, by six o'clock the whole monstrous pile had become a funeral pyre for perhaps as many as eighty people trapped inside.

Editor George Swank, who had been watching everything from his window at the *Tribune* office, wrote that it burned "with all the fury of the hell you read about—cremation alive in your own home, perhaps a mile from its foundation; dear ones slowly consumed before your eyes, and the same fate yours a moment later."

By ten o'clock the light from the flames across the lower half of town was bright enough to read a newspaper by.

-2-

The water in front of the Heiser store had been knee-deep since early in the afternoon, which was a record for that part of town. In the other floods over the years there had never been any water at all so far up on Washington Street.

People had been coming in and out of the store most of the morning joking about the weather, buying this and that to tide them through the day. The floor was slick with mud from their boots, and the close, warm air inside the place smelled of tobacco and wet wool. George Heiser, wearing his usual old sweater, was too busy taking care of customers to pay much attention to what was going on outside.

But by early afternoon, with the street out front under two feet of water, hardly anyone was about, and the Heiser family was left more or less to itself. A few visitors dropped in, family friends, and an occasional customer. Mrs. Lorentz, from Kernville, sat visiting with Mathilde Heiser upstairs. She had come by alone, without her husband, who was the town's weatherman, and, no doubt, a busy man that day.

Sometime near four o'clock George Heiser had sent his son, Victor, out to the barn to see about the horses. The animals had been tied in their stalls, and George, worried that they might strangle if the water should get any higher, wanted them unfastened.

The barn, like the store front, was a recent addition for the Heisers. It had a bright-red tin roof and looked even bigger than it was, standing, as it did, upon higher ground at the rear of their lot. To get back to it, Victor had left his shoes and socks behind and, with a pair of shorts on, went wading across through the pelting rain. It had taken him only a few minutes to see to the horses and he was on his way out the door when he heard the noise.

Terrified, he froze in the doorway. The roar kept getting

louder and louder, and every few seconds he heard tremendous crashes. He looked across at the house and in the second-story window saw his father motioning to him to get back into the barn and up the stairs. Just a few weeks earlier he and his father had cut a trap door through the barn roof, because his father had thought "it might be a good idea."

The boy was through the door and onto the roof in a matter of seconds. Once there he could see across the top of the house, and on the other side, no more than two blocks away, was the source of all the racket. He could see no water, only an immense wall of rubbish, dark and squirming with rooftops, huge roots, and planks. It was coming at him very fast, ripping through Portage and Center streets. When it hit Washington Street, he saw his home crushed like an orange crate and swallowed up.

In the same instant the barn was wrenched from its footings and began to roll like a barrel, over and over. Running, stumbling, crawling hand over hand, clawing at tin and wood, Victor somehow managed to keep on top. Then he saw the house of their neighbor, Mrs. Fenn, loom up in front. The barn was being driven straight for it. At the precise moment of impact, he jumped, landing on the roof of the house just as the walls of the house began to give in and the whole roof started plunging downward.

He clambered up the steep pitch of the roof, fighting to keep his balance. The noise was deafening and still he saw no water. Everything about him was cracking and splitting, and the air was filled with flying boards and broken glass. It was more like being in the middle of an explosion than anything else.

With the house and roof falling away beneath him, he caught hold of still another house that had jammed in on one side. Grabbing on to the eaves, he hung there, dangling, his feet swinging back and forth, reaching out, trying to get a toe hold. But there was none. All he could do was hang and swing. For years after he would have recurring nightmares in which it was happening to him all over again. If he let go he was finished. But in the end, he knew, he would have to let go. His fingernails dug deep into the water-soaked shingles. Shooting pains ran through his hands and down his wrists.

Then his grip gave out and he fell, backwards, sickeningly, through the wet, filthy air, and slammed down on a big piece of red roof from the new barn. And now, for the first time, he saw water; he was bumping across it, lying on his stomach, hanging on to the roof with every bit of strength left in him, riding with the wave as it smashed across Johnstown.

The things he heard and saw in the next moments would be remembered later only as a gray, hideous blur, except for one split-second glimpse which would stick in his mind for the rest of his life.

He saw the whole Mussante family sailing by on what appeared to be a barn floor. Mussante was a fruit dealer on Washington Street, a small, dark Italian with a drooping mustache, who had been in Johnstown now perhaps three years. He had had a pushcart at first, then opened the little place not far from the Heiser store. Victor knew him well, and his wife and two children. Now there they were speeding by with a Saratoga trunk open beside them, and every one of them busy packing things into it. And then a mass of wreckage heaved up out of the water and crushed them.

But he had no time to think more about them or anything else. He was heading for a mound of wreckage lodged between the Methodist Church and a three-story brick building on the other side of where Locust Street had been. The next thing he knew he was part of the jam. His roof had catapulted in amongst it, and there, as trees and beams shot up on one side or crashed down on the other, he went leaping back and forth, ducking and dodging, trying desperately to keep his footing, while more and more debris kept booming into the jam.

Then, suddenly, a freight car reared up over his head. It looked like the biggest thing he had ever seen in his life. And this time he knew there could be no jumping out of the way.

But just as it was about to crash on top of him, the brick building beside him broke apart, and his raft, as he would describe it later, "shot out from beneath the freight car like a bullet from a gun."

Now he was out onto comparatively open water, rushing across a clear space which he judged to be approximately where the park had been. He was moving at a rapid clip, but there seemed far less danger, and he took some time to look about.

There were people struggling and dying everywhere around him. Every so often a familiar face would flash by. There was Mrs.

Fenn, fat and awkward, balanced precariously on a tar barrel, well doused with its contents, and trying, pathetically, to stay afloat. Then he saw the young Negro who worked for Dr. Lee, down on his knees praying atop his employer's roof, stark naked, shivering, and beseeching the Lord in a loud voice to have mercy on his soul.

Like the Mussante family, they were suddenly here and gone like faces in nightmares, or some sort of grotesque comedy, as un-real and as unbelievable as everything else that was happening. And there was nothing he could do for them, or anybody else.

He was heading across town toward the Stony Creek. As near as he could reckon later, he passed right by where Horace Rose's house had stood, then crossed Main and sailed over the Morrell lot, and perhaps directly over where the Morrell greenhouse had been. Almost immediately after that, about the time he was crossing Lincoln Street, he got caught by the backcurrent.

Until then he had been keeping his eyes on the mountainside, which looked almost close enough to reach out and touch, and on the stone bridge. Both places looked to be possible landings, and either one would do as well as the other.

But now his course changed sharply, from due west to due south. The current grabbed his raft and sent it racing across the Stony Creek a half mile or so, over into the Kernville section, and it was here that his voyage ended.

"I passed by a two-and-a-half-story brick dwelling which was still remaining on its foundations. Since my speed as I went up this second valley was about that of a subway train slowing for a stop, I was able to hop to the roof and join a small group of people already stranded there."

When he had been standing on the roof of his father's barn, looking across the housetops at the avalanche bearing down on Johnstown, he had taken his watch out of his pocket to look at the time. It was a big silver watch with a fancy-etched cover, which had been his fourteenth birthday present from his father. He had snapped it open, because, as he would say later, "I wanted to see just how long it was going to take for me to get from this world over into the next one."

Now, on the rooftop in Kernville, realizing that he had per-haps a very good chance of staying on a little longer in this world, he pulled out the watch a second time.

Amazingly enough, it was still running, and he discovered with astonishment that everything that had happened since he had seen his home vanish had taken place in less than ten minutes.

Agnes Chapman had watched her husband walk to the front door in his bedroom slippers about four o'clock, open it, peer out, and turn around looking, as she told it later on, "pale and affrighted." The Reverend had just seen a boxcar with a man standing on top roll down the pavement in front of the parsonage. As he passed under the tree in the Chapmans' yard, the man had caught hold of a limb and swung himself up onto the roof of the front porch, from which he stepped through the second-story window directly over the Reverend's head.

The man was the ticket agent from the B & O station, across Washington Street from the Heiser store. Upon hearing the commotion up the valley, he had climbed on top of the car to see what was going on. Then the car had started running with what must have been a small but powerful current preceding the main wave. It swept the car down Franklin, across Locust, too fast for the man to do anything but hang on until he was within reach of the Chapmans' tree.

The whole scene meant only one thing to the Reverend. The reservoir had broken. He shouted for everyone to run for the attic.

Agnes Chapman, with her seven-year-old granddaughter, Nellie, Mrs. Brinker (their neighbor from across the park), Mr. Parker, and Lizzie, the cook, all made a dash up the front stairs, while Chapman ran to the study to shut off the gas fire. As he turned to go back out to the hall, he saw the front door burst open and a huge wave rush in. He ran for the kitchen and scrambled up the back stairs. A few seconds more and he would have been swept against the ceiling and drowned. The water was up the stairs and into the second floor almost instantly.

By now the whole family was in the attic, along with the B & O ticket agent and two other young men who had jumped through an open window from a whirling roof.

"We all stood there in the middle of the floor, waiting our turn to be swept away, and expecting every minute to be drowned." Mrs. Chapman said. "When our porches were torn loose, and the two bookcases fell over, the noise led us to think the house was going to pieces."

The noise everywhere was so awful they had to shout to hear one another. Outside other buildings were scraping and grinding against theirs, or crashing in heaps, and the thunder of the water kept on for what seemed an eternity.

"We knew . . . that many of our fellow citizens were perishing, and feared that there could be no escape for us," the Reverend Chapman wrote later. "I think none was afraid to meet God, but we all felt willing to put it off until a more propitious time . . ."

About then a man Chapman thought to be "an Arabian" came bounding through the window, clad only in underdrawers and a vest. He was drenching wet, shaking with cold and terror, and kept shouting at them, "Fader, Mudder. Tronk! Tronk! Two, tree hooner tollar, two, tree hooner tollar."

"I think he wanted to tell us he had lost his trunk with two or three hundred dollars he had saved to bring his mother and father over here," Chapman later explained.

The man got right down on his knees and started praying over a string of beads with such frenzy that the Reverend had to quiet him down, as he "excited and alarmed the ladies."

But despite everything happening outside, the parsonage appeared to be holding on. And when the roar began to die off, Chapman went to the window to take a look. It was, he wrote afterward, "a scene of utter desolation." With darkness closing down on the valley and the rain still falling, his visibility was quite limited. Still, he could make out the tall chimneys and gables of Dr. Lowman's house across the park, poking above what looked to him like a lake spread over the town at a depth of maybe thirty feet. There was not a sign of any of the other houses that had been on the park, but over on the left, where Main Street had been, he could see the dim silhouettes of the bank, Alma Hall, which was the Odd Fellows new building, and the Presbyterian Church sticking up out of the dark water. There were no lights anywhere and no people. "Everyone is dead," Chapman thought to himself.

Mrs. Brinker asked him to look to see if her house was still standing. When he said it was not, the others did what they could to console her. The room grew steadily darker, and from outside came more sounds of houses cracking up and going down under the terrible weight of the water.

The Hulbert House had been the finest hotel in town. It was not so large as the Merchants' Hotel on Main, but it was newer and fitted out "with all the latest wrinkles" as one paper of the day put it. Drummers made up most of the trade, and things were arranged to suit them. Breakfast was served early, dinner at noon (a custom most big-city hotels had long since abandoned), and like the other chief hotels in town, each of its rooms had a long extension table where the salesmen could display their wares. "Through some open door we can always see one piled high with samples of the latest fashions as adulterated for the provincial market," wrote a visitor from New York. It was also, for some strange reason, the only hotel in town without a bar.

Located on Clinton Street, three doors from Main on the east side of the street, it was all brick and four stories tall. Earlier that morning it had looked to quite a number of people like one of the safest places in town.

For example, Jeremiah Smith, a stonemason who lived in a small frame house over on Stony Creek Street, brought his wife and three children (nine-year-old Florence, seven-year-old Frank, and a four-month-old baby) across town through the rain to the safety of the Hulbert House. How long Smith stayed on with them is not known, but the evidence is he soon went back home again. In any case, he and his house survived the flood. His wife and children were crushed to death when the Hulbert House collapsed almost the instant it was hit by the flood.

In all there were sixty people inside the building by four o'clock in the afternoon. Only nine of them got out alive.

"Strange as it may seem, we were discussing the possibility of the dam breaking only a few hours before it really did," one of the survivors, a G. B. Hartley of Philadelphia, was later quoted.

"We were sitting in the office shortly after dinner. Everyone laughed at the idea of the dam giving way. No one had the slightest fear of such a catastrophe."

As the afternoon passed, Hartley moved to the second-floor parlor. He was sitting there talking to a Miss Carrie Richards, Charles Butler of the Cambria Iron Company, and Walter Benford, brother of the proprietor, when they heard shouting in the streets, immediately followed by loud crashes.

"At first sound," Hartley said, "we all rushed from the room panic-stricken. Why it was I do not know, but we ran for the stairs. Mr. Butler took Miss Richards' hand. She called to me, and I took hold of her other hand. Then we started up the stairs. Mr. Benford did not go with us, but instead ran downstairs where his brother had an office. The scene in the hotel is beyond imagination or description.

"Chambermaids ran screaming through the halls, beating their hands together and uttering wild cries to heaven for safety. Frightened guests rushed about not knowing what to do nor what was coming. Up the stairs we leapt. Somewhere, I do not know when or how it was, I lost my hold of Miss Richards' hand. I really cannot tell what I did, I was so excited. I still rushed up the stairs and thought Miss Richards and Mr. Benford were just behind and I had reached the top flight of stairs and just between the third and fourth floors, when a terrific crash came. Instantly I was pinned by broken boards and debris . . ."

Hartley then looked up and saw that the building's big mansard roof had been lifted right off and he was looking at nothing more than a sullen sky. In what must have been no more than thirty seconds or so, he managed to scramble out from under the debris and climb onto the roof, which was floating to the side of the crumbling hotel.

F. A. Benford, proprietor of the house, was already on the roof, along with his brother Walter, a traveling salesman from Strawbridge & Clothier named Herbert Galager, and two chambermaids, one of whom had a dislocated shoulder. The roof floated off with the current. The rest of the building just disappeared; the walls fell in and it was gone.

Gertrude Quinn was the six-year-old daughter of James Quinn, who, with his brother-in-law, Andrew Foster, ran Geis, Foster and Quinn; Dry Goods and Notions, which stood diagonally across Clinton from the Hulbert House. The two of them, Gertrude would later say, looked like the Smith Brothers on the cough-drop box.

James Quinn was one of the few prominent men in Johnstown

who had been noticeably concerned about the dam since early that morning. He had been to the lake several times over the years and had a clear idea of the volume of water there. If the dam should let go, he had said, not a house in town would be left standing.

The Quinns lived in one of Johnstown's show places, a three-story, red-brick Queen Anne house newly built at the corner of Jackson and Main. It was surrounded by an iron fence and stood well up off the street, perfectly safe, it was to be assumed, from even the worst spring floods. There were fruit trees and a flower garden in the front yard, a kitchen garden, a barn with one cow and some ducks out back. Inside, everything was the latest—plumbing, icebox, organ, piano, Arab scarves, Brussels carpets, a marble clock from Germany on the mantel.

Besides Gertrude, there were six other children in the family. Vincent, who was sixteen, was the oldest. Helen, Lalia, and Rosemary came next; then Gertrude, Marie, and Tom, who was only a few months old. Rosina Quinn, their mother, was the daughter of old John Geis, who had started the store back in canal days, soon after he arrived from Bavaria. She had worked in the business herself before marrying and was later teased for having five of her seven children in July, which, as everyone knew, was the slow season for dry goods.

Then there was Libby Hipp, the eighteen-year-old German nursegirl, Gertrude's Aunt Abbie (Mrs. Geis), and her infant son, Richard. Aunt Abbie, who was probably no more than twenty-eight years old and a woman of exceptional beauty, had come east for her health from her home in Salina, Kansas. She had had three children in a very short time and needed rest.

James Quinn was most definitely head of the household. He was a trim, bookish man who had been an officer in the cavalry during the war and still held himself in a like manner. He was President of the Electric Light Company, a member of the school board, and, along with Cyrus Elder, Dr. Lowman, and George Swank, he was one of the trustees of the Johnstown Savings Bank. As a boy he had been taken by his father, a construction worker, to ask for a job in the Cambria mills but had been turned down because he looked too scared—for which he would be forever thankful. For a while before the war he had toyed with the idea of becoming an artist, and one of his early efforts, *Rebecca at the Well*, done in house paints, hung in the third floor of the new house on Jackson Street. (Later on, his wife would tell him, "The flood wasn't so bad, when you realize we got rid of *Rebecca* so gracefully.")

At home he was quite exacting about the use of the English language, abhorring slang and insisting on proper diction. He liked cigars. He was quiet, dignified, a strong Republican, and a good Catholic.

The advertisements he was placing in the *Tribune* that spring let it be known that Foster and Quinn were offering the finest in Hamburg embroideries, Spanish laces, Marseilles quilts, and "new French sateens." But the store also dealt in carpets, umbrellas, hat-pins, hairpins, flannel drawers, striped calico dresses, pearl buttons, black hose, bolsters, and pillowcases.

"I cannot separate thoughts of parents, brothers, sisters, or home from our store," Gertrude would say later. "When we went there, we became personages . . . the clerks, vying with one an-other for our attention, were always doing thoughtful little things for us."

The place was big and brightly lighted, with people coming and going, exchanging news and gossip. For the children it was all a grand show, from which they took home strings of stray beads or buttons or some other trinket.

For Foster and Quinn (father-in-law Geis had long since retired), the place represented an investment of about $60,000 and provided a very good living.

On the morning of the 31st, James Quinn had gone to the store early to supervise the moving of goods to higher levels. Before leav-ing home he had told everyone to stay inside. One of his children, Marie, was already sick with measles, and he did not want the others out in the rain catching cold. He did, however, allow young Vincent to come along with him downtown to lend a hand.

At noon, when he had returned for dinner, the water had been up to his curbstone. He had been restless and worried through the meal, talking about the water rising in the streets and his lack of confidence in the South Fork dam.

A few days before, he and his wife and the infant, Tom, and Lalia had gone to Scottdale for a christening, and Mrs. Quinn and the two children had stayed on to visit with her sister. Now Aunt Abbie and Libby Hipp were more or less running things, and he was doing his best to make sure they understood the seriousness of the situation.

"James, you are too anxious," his sister-in-law said. "This big house could never go."

In recalling the day years afterward, Gertrude felt sure that her father was so worried that he would have moved them all to the hill that morning, even though he had no special place to take them, if it had not been for Marie. He was afraid of the effect the light might have on her eyes.

After dinner he had gone back to the store, and Gertrude slipped out onto the front porch where she began dangling her feet in the water, which, by now, covered the yard just deep enough for the ducks to sport about among the flowers. Everyone who survived the flood would carry some especially vivid mental picture of how things had looked just before the great wave struck; for this child it would be the sight of those ducks, and purple pansies floating face up, like lily pads, in the yellow water.

Shortly before four Gertrude's father suddenly appeared in front of her. He took her with one hand, with the other gave her a couple of quick spanks for disobeying his order to stay inside, and hurried her through the door.

"Then he gave me a lecture on obedience, wet feet, and our perilous position; he said he had come to take us to the hill and that we were delayed because my shoes and stockings had to be changed again. He was smoking a cigar while the nurse was changing my clothes. Then he went to the door to toss off the ashes."

It was then that he saw the dark mist and heard the sound of the wave coming. He rushed back inside, shouting, "Run for your lives. Follow me straight to the hill."

Someone screamed to him about the baby with the measles. He leaped up the stairs and in no more than a minute was back down with Marie wrapped in a blanket, his face white and terrified-looking.

"Follow me," he said. "Don't go back for anything. Don't go back for anything." Everyone started out the door except Vincent. Just where he was no one knew. Helen and Rosemary ran on either side of their father, holding on to his elbows as he carried the baby. When they got to the street the water was nearly to Rosemary's chin, but she kept going, and kept trying to balance the umbrella she had somehow managed to bring along. The hill was at most only a hundred yards away. All they had to do was get two short blocks to the end of Main and they would be safe.

James Quinn started running, confident that everyone was with him. But Aunt Abbie, who was carrying her baby, and Libby Hipp, who had Gertrude in her arms, had turned back.

When she reached the top of the steps that led from the yard down to the street, Aunt Abbie had had second thoughts.

"I don't like to put my feet in that dirty water," Gertrude would remember her saying. Libby said she would do whatever Aunt Abbie thought best, so they started back into the house.

"Well, I kicked and scratched and bit her, and gave her a terrible time, because I wanted to be with my father," Gertrude said later. How the two women, each with a child, ever got to the third floor as fast as they did was something she was never quite able to figure out. Once there, they went to the front window, opened it, and looked down into the street. Gertrude described the scene as looking "like the Day of Judgment I had seen as a little girl in Bible histories," with crowds of people running, screaming, dragging children, struggling to keep their feet in the water.

Her father meanwhile had reached dry land on the hill, and turning around saw no signs of the rest of his family among the faces pushing past him. He grabbed hold of a big butcher boy named Kurtz, gave him Marie, told him to watch out for the other two girls, and started back to the house.

But he had gone only a short way when he saw the wave, almost on top of him, demolishing everything, and he knew he could never make it. There was a split second of indecision, then he turned back to the hill, running with all his might as the water surged along the street after him. In the last few seconds, fighting the current around him that kept getting deeper and faster every second, he reached the hillside just as the wave pounded by below.

Looking behind he saw his house rock back and forth, then lunge sideways, topple over, and disappear.

Gertrude never saw the wave. The sight of the crowds jamming through the street had so terrified her aunt and Libby Hipp that they had pulled back from the window, horrified, dragging her with them into an open cupboard.

"Libby, this is the end of the world, we will all die together," Aunt Abbie sobbed, and dropped to her knees and began praying hysterically, "Jesus, Mary, and Joseph, Have mercy on us, oh, God . . ."

Gertrude started screaming and jumping up and down, calling "Papa, Papa, Papa," as fast as she could get it out.

The cupboard was in what was the dining room of an elaborate playhouse built across the entire front end of the third floor. There was nothing like it anywhere else in town, the whole place having been fitted out and furnished by Quinn's store. There was a long center hall and a beautifully furnished parlor at one end and little bedrooms with doll beds, bureaus, washstands, and ingrain carpets on the floors. The dining room had a painted table, chairs, sideboard with tiny dishes, hand-hemmed tablecloths, napkins, and silverware.

From where she crouched in the back of the cupboard, Gertrude could see across the dining room into a miniature kitchen with its own table and chairs, handmade iron stove, and, on one wall, a whole set of iron cooking utensils hanging on little hooks. Libby Hipp was holding her close, crying and trembling.

Then the big house gave a violent shudder. Gertrude saw the tiny pots and pans begin to sway and dance. Suddenly plaster dust came down. The walls began to break up. Then, at her aunt's feet, she saw the floor boards burst open and up gushed a fountain of yellow water.

"And these boards were jagged . . . and I looked at my aunt, and they didn't say a word then. All the praying stopped, and they gasped, and looked down like this, and were gone, immediately gone."

She felt herself falling and reaching out for something to grab on to and trying as best she could to stay afloat.

"I kept paddling and grabbing and spitting and spitting and trying to keep the sticks and dirt and this horrible water out of my mouth."

Somehow she managed to crawl out of a hole in the roof or wall, she never knew which. All she saw was a glimmer of light, and she scrambled with all her strength to get to it, up what must have been the lath on part of the house underneath one of the gables. She got through the opening, never knowing what had become of her aunt, Libby, or her baby cousin. Within seconds the whole house was gone and everyone in it.

The next thing she knew, Gertrude was whirling about on top of a muddy mattress that was being buoyed up by debris but that kept tilting back and forth as she struggled to get her balance. She screamed for help. Then a dead horse slammed against her raft, pitching one end of it up into the air and nearly knocking her off. She hung on for dear life, until a tree swung by, snagging the horse in its branches before it plunged off with the current in another direction, the dead animal bobbing up and down, up and down, in and out of the water, like a gigantic, gruesome rocking horse.

Weak and shivering with cold, she lay down on the mattress, realizing for the first time that all her clothes had been torn off except for her underwear. Night was coming on and she was terribly frightened. She started praying in German, which was the only way she had been taught to pray.

A small white house went sailing by, almost running her down. She called out to the one man who was riding on top, straddling the peak of the roof and hugging the chimney with both arms. But he ignored her, or perhaps never heard her, and passed right by.

"You terrible man," she shouted after him. "I'll never help you."

Then a long roof, which may have been what was left of the Arcade Building, came plowing toward her, looking as big as a steamboat and loaded down with perhaps twenty people. She called out to them, begging someone to save her. One man started up, but the others seemed determined to stop him. They held on to him and there was an endless moment of talk back and forth between them as he kept pulling to get free.

Then he pushed loose and jumped into the current. His head bobbed up, then went under again. Several times more he came up and went under. Gertrude kept screaming for him to swim to her. Then he was heaving himself over the side of her raft, and the two of them headed off downstream, Gertrude nearly strangling him as she clung to his neck.

The big roof in the meantime had gone careening on until it hit what must have been a whirlpool in the current and began spinning round and round. Then, quite suddenly, it struck something and went down, carrying at least half its passengers with it.

Gertrude's new companion was a powerful, square-jawed millworker named Maxwell McAchren, who looked like John L. Sullivan. How far she had traveled by the time he climbed aboard the mattress, she was never able to figure out for certain. But later on she would describe seeing many flags at one point along the way, which suggests that she went as far up the Stony Creek as Sandy Vale Cemetery, where the Memorial Day flags could have been visible floating about in the water. Sandy Vale is roughly two miles from where the Quinn house had been, and when Maxwell McAchren joined her, she had come all the way back down again and was drifting with the tide near Bedford Street in the direction of the stone bridge.

On a hillside, close by to the right, two men were leaning out of the window of a small white building, using long poles to carry on their own rescue operation. They tried to reach out to the raft, but the distance was too great. Then one of them called out, "Throw that baby over here."

McAchren shouted back, "Do you think you can catch her?"

"We can try," they answered.

The child came flying through the air across about ten to fifteen feet of water and landed in the arms of Mr. Henry Koch, proprietor of Koch House, a small hotel and saloon (mostly saloon) on Bedford Street. The other man in the room with him was George Skinner, a Negro porter, who had been holding Koch by the legs when he made the catch. The men stripped Gertrude of her wet underclothes, wrapped her in a blanket, and put her on a cot. Later she was picked up and carried to the hill, so bundled up in the warm blanket that she could not see out, nor could anyone see in very well.

Every so often she could hear someone saying, "What have you got there?" And the answer came back, "A little girl we rescued." Then she could hear people gathering around and saying, "Let's have a look." Off would come part of the blanket in front of her face and she would look out at big, close-up faces looking in. Heads would shake. "Don't know her," they would say, and again the blanket would come over her face and on they would climb.

Gertrude never found out who it was who carried her up the hill, but he eventually deposited her with a family named Metz, who lived in a frame tenement also occupied by five other families. The place looked like paradise to her, but she was still so terrified that she was unable to say a word as the Metz children, neighbors, and people in off the street jammed into the kitchen to look at her as she lay wrapped now in a pair of red-flannel underwear with Mason jars full of hot water packed all around her.

Later, she was put to bed upstairs, but exhausted as she was she was unable to sleep. In the room with her were three other refugees from the disaster, grown women by the name of Bowser, who kept getting up and going to the window, where she could hear them gasping and whispering among themselves. After a while Gertrude slipped quietly out of bed and across the dark room. Outside the window, down below where the city had been, she could now see only firelight reflecting on water. It looked, as she said later, for all the world like ships burning at sea.

The Reverend Dr. David Beale, pastor of the Presbyterian Church on Main Street, was one of the several hundred people crowded into the cavernous, pitch-dark rooms of the second, third, and fourth floors of Alma Hall. The building was on Main, five doors up from Dr. Beale's church and directly across the street from the park. It was the tallest, largest structure in Johnstown.

Dr. Beale had been at home that afternoon, in the Lincoln Street parsonage, which stood directly behind his church, and like his good friend and neighbor the Reverend Chapman, he had been working on his sermon for Sunday. About four he had gone into the parlor to help take up the carpet. Then all at once the house was struck, and in the next few seconds he snatched up the family Bible, his wife turned off the gas, his daughter grabbed the canary cage, and they and several neighbors who had dropped by earlier all dashed up the front stairs. By the time they reached the second floor the water was up to their waists, and a hat rack was driven against Beale's back with such force that it nearly knocked him under. As they reached the third floor a man washed in through the window.

"Who are you? Where are you from?" Beale shouted.

"Woodvale," the man gasped. He had been carried on a roof a mile and a quarter.

Expecting at any moment "to be present with the Lord," Beale led the group in a prayer and read aloud from the Bible, his voice straining against the noise of the flood:

"God is our refuge and strength, a very present help in trouble.

"Therefore will not we fear, though the earth be removed, and though the mountains be carried into the midst of the sea . . ."

Outside twisted wreckage, tank cars, freight cars, and what appeared to be every house in sight went tumbling past the window. The Reverend Beale saw J. Q. Benshoff, Johnstown's leading bookseller go by, Mrs. John Fulton and her daughter, and two small children clinging to a roof, both of them nearly naked. For blocks around every building appeared to have been obliterated.

There were ten people in the Beale attic, counting the newcomer from Woodvale. Soon after, Beale helped save Captain A. N. Hart, his wife, sister, and two small sons by pulling them in through the window. But there was considerable doubt as to how long the frame parsonage could last, and especially when, after a shifting and quieting of the current, the wreckage which had been shoved against the west side of the house began slowly drifting off and the whole building started to jerk and tremble.

A decision was made to try, before nightfall, to walk across the flood—over the debris—to Alma Hall, which was the equivalent of about one block away. Captain Hart went out the window first on the end of a rope, tested a roof that was floating below, found it stable enough, and the other fourteen, plus the Beales' dog, a terrier called "Guess," followed after. The Reverend Beale was the last one out. Then they started off, picking their way over tree trunks, timbers, stepping from one moving house to another, climbing up the sides of roofs sometimes so steep that part of the group on one side would be out of sight from those on the other side, then jumping across sudden spaces of dark water or bridging them with stray planks. At one point one of the girls lost her balance and fell in, vanishing from sight except for her hair which floated on the surface. She was rescued by pulling her back on some long boards, and everyone continued on.

By now, very near dark, the city was one huge, vile-looking lake anywhere from ten to thirty feet deep, much of it crusted over with a grinding pack of wreckage, across which now other groups of tiny figures, barely visible in the fading light, could be seen groping their way toward the hills or the few buildings still left standing.

About the time the Reverend Beale and his party had climbed out of the parsonage, a break had occurred in the railroad embankment to the right of the stone bridge, between the bridge and the depot, and the water began raging through just as though it were a spillway. House after house had plunged through the break like boats running the rapids, many of them loaded with people, dozens to be dashed to pieces when they hit Cambria City below.

But there were some people who, one way or another, survived the trip to be fished out by rescue teams farther along the river. One of them was Maxwell McAchren, who, after throwing little Gertrude to safety, sailed on toward the bridge in time to be sucked through the break. He wound up riding the mattress straight through Cambria City at the time when a good part of it was being destroyed, past the Cambria works as they were being pounded by the water, and on down the Conemaugh four miles before he was finally pulled to shore by a crowd of men with ropes.

With the break in the embankment the level of the water over the city began to go down, but only slowly, since the Little Conemaugh and the Stony Creek were still pouring in immense quantities of water, mud, and debris. So as night began, those buildings which had somehow held up against everything so far were still withstanding as much as twenty feet of water, and very often they had had several hundred tons of wreckage dumped against them. For those who would manage to get inside them, the long night ahead would be by far the most agonizing part of the whole ordeal.

When the Reverend Beale's group finally reached Alma Hall, there were already close to 200 people inside. At the Dean Canan house there were 60 people in the attic. At least 51 people were in the attic of Dr. Walters' house on Vine Street. The Fred Krebs house had 125 people inside before the night ended. Nearly 200 people were in the upper floors of the Union Street School, more than 100 on top of the Wolfe Building, and at the Morrell house (by then it had been converted into the Morrell Institute, a vocational training school for the Iron Company) there were 175 people. And over at Dr. Swan's tall brick house at the corner of Vine and Stony Creek streets there were close to 90 people, including Horace Rose, who lay stretched out on the floor with a dislocated shoulder, a broken collarbone, several crushed ribs, and half of his face ripped open.

There had been a few minutes after the flood had fallen on his part of Main Street during which Rose had been at his window, almost hypnotized by the scene outside. He had seen John Dibert's house squashed like a paper bag. Another brick house fell with a crash. A large frame building directly across the street had lifted up and charged right for him. Then there had been a horrible noise, he had felt himself falling, and all was dark.

"A moment later I felt the press of a heavy shock, a sense of excruciating pain . . . the thought came upon me that I was being crushed to death . . ."

His whole right side had been caved in by falling timbers, and he was powerless to free himself. He had heard his youngest son calling for help but had been unable to do anything for him. He had seen his daughter, June, rise up out of the water, then, almost immediately, sink back out of sight. From out of nowhere a small boy had appeared among the chaos and told him his wife had drowned. Then another stranger, this one a young man, seemed, Rose said later, to shoot out of the debris. Rose told him to go help his wife and daughter. The man, Rose learned afterward, was a Pittsburgh dentist named Phillips, and in a few frantic minutes he managed to free Mrs. Rose from the timbers that had fallen on her. Then the majority of the Rose household—Rose, his wife and daughter (she too had been rescued somehow), two of his sons, one maid, "the strange boy," as Rose called him, and an elderly lady who had been pulled off a floating shutter by one of his sons—were all together on a single stout roof which chanced by at the very moment when the last of their house was disappearing in the tide. The roof had been heading toward the stone bridge. But, Rose wrote later, "Scarcely was the complement of passengers complete, when the current turned, and our ship was driven with terrific velocity directly up the channel of the Stony Creek . . ."

Then for several more hours they had floated about, sometimes wallowing in dead water, other times rushing rapidly back over a course they had just completed. And through it all Rose lay helpless, in terrible pain, and shaking with severe chills as the cold rain beat down.

They had seen the spire of St. John's Catholic Church catch fire, which according to most accounts had happened about eight o'clock, and had watched the flames leap clear to the cross on top before the whole thing toppled and fell into the water. At another point they had been becalmed within perhaps a hundred feet of where Rose's office had been on Franklin Street, and listened to the ringing of the ponderous bell in the town clock. The clock was in the steeple of the Lutheran Church, and somehow or other its mechanism was still functioning the same as ever. Through the rest of the night, despite everything, every hour on the hour, it bonged away. The sound had a powerful effect on everyone who heard it.

Then by another sudden change in the surface currents, the roof had been driven off over the main channel of the Stony Creek, where by now the current was again heading downstream. They were carried a hundred yards or more before the roof lodged against the side of the Swan house.

Rose was lifted from the roof and through a window. From then until morning he lay listening to buildings breaking up somewhere out in the night and watching the light from the fire at the bridge play across the walls and ceiling.

For the Reverend Beale and the others inside Alma Hall there had been an immediate fear of fire and what might happen if panic should break out among so many people waiting in the dark. An Alma Hall government had been set up, with Beale and Captain Hart each put in charge of one of the floors. Some whiskey was confiscated, and the use of matches was strictly forbidden because of the likelihood of a natural-gas leak in the basement. A count was made to see how many there were (it came to 264), and the Reverend once again led a prayer.

James Walters, a lawyer, was named director of the building. Walters had made one of the day's most extraordinary voyages, having been swept from his home on Walnut Street on top of a roof which took him spinning across town until he smashed into the side of Alma Hall, flew headlong through a window, and landed square in the middle of his own office.

The fourth member of the governing body was the only physician in the building, Dr. William Matthews, who spent the entire night tending to the wounded, without sleep or rest, despite the fact that he had two broken ribs.

In the Reverend Beale's words, it was a "night of indescribable horrors." The only light was the faint, eerie glow from the fires outside. Up near the long front windows that earlier in the day had looked down into the green treetops of the park, the light was bright enough to recognize a nearby face; but farther back in the deep, high-ceilinged rooms it was nearly pitch-black, and on the stairways between floors there was no light at all.

Nearly everyone was wringing wet, filthy, and suffering from the cold. A number of people had most of their clothes torn off. There was no food and no water. There were no blankets, no dry clothes, and no medical supplies. The injured lay shivering in the dark. The rooms were filled with their moaning, with the crying of scared, hungry children, and with a lot of fervent praying.

Outside they could hear the rush of the rain and faint calls for help, a sudden scream, and every now and then the unearthly howling of dogs and other animals, which to many people was the most frightful sound of all.

Nor was there any assurance whatsoever that the whole enormous building would not go the way of so many others and crack apart and bury them all under tons of brick and plaster and falling timbers. Everyone was asked to move about as little as possible. According to Beale, "the expressed opinion of the contractors present" was that the building would not last the night.

People began thinking about whether their own corpses would be recognizable or where they might be buried, if ever their bodies were found. The suspense was unbearable, and it kept on, hour after hour. It seemed morning would never come.

But Alma Hall stood through the night, as did the Presbyterian Church and its parsonage, Dr. Lowman's house, where a small crowd had gathered in the top floor, and the Methodist parsonage, where the Chapmans and their assorted guests huddled together in the numbing cold praying for morning. The buildings survived because they were on the lee side of the big, stone Methodist Church. Standing as it did, at the corner of Franklin and Locust, on the northeastern corner of the park, the church was one of the first sizable buildings in town to be struck by the wave. Not only had it held, but it had split the wave and so served as a shield for buildings directly in line behind it. (One tale to come out of Alma Hall later on told of a voice in the dark saying, "We've been saved by the Methodist Church," whereupon another voice answered back, "Only the Catholic Church can save!")

Elsewhere in the night the story was quite different. Buildings caved in or caught fire and burned to the water line. The St. John's fire was the biggest and most spectacular, but there were fires among several houses close by; the Keystone Hotel caught fire and there were one or two small fires over in Kernville.

And aside from the many large groups of people gathered inside Alma Hall, Dr. Swan's house, or the other buildings that were still standing, there were any number of smaller groups of four to six, or even one to two, people who spent the night inside their own tiny attics or atop the roofs of little houses that bobbed about with the current. Some were closed in under roof beams, with no windows to look out or escape through; they were still alive, but trapped, and with no way of knowing what might happen next.

At least one family had jumped into a large bed when the water rushed up their stairs. The bed was borne clear to the ceiling by the water, and the family stayed there, floating inside their own house through the remainder of the night.

Another family named Williams had their house split in half at the bridge, then went floating up the Stony Creek in what was left of the attic. In the darkness that night Mrs. Williams gave birth to a baby boy; and the family stayed there until morning, soaked, freezing cold, the baby wrapped in a shawl.

Scores of others floated on rooftops or freight cars or half-submerged debris, without any protection from the pouring rain. A Mrs. Jacob Malzi hung on to the eaves of a house all night, up to her waist in water. A Miss Minnie Chambers had climbed inside a freight car which had been carried through the cut near the bridge and smashed to pieces against the roof of the Cambria works, where she, miraculously still alive, spent the night holding on to a small pipe that stuck up through the roof. James Shumaker lay half-unconscious across a heap of drifting wreckage all night, his face and arms badly torn and nearly blinded in both eyes by sand and lime.

Several people spent the night in trees, hanging on with the water lapping about below, never daring to close their eyes, even for a few moments, for fear they might fall asleep, lose their grip, and drop into the black current. Jacob Horner and his family of eight spent all night in a tree; so did Reuben Bensen and Mrs. Ann Buck, who was eighty years old, and Mrs. John Burket, who had had every bit of her clothing ripped from her back by the flood.

Of the great many people who were lucky enough to get to dry land, there were a number who were in such a state of shock and fear that they just started walking, stopping for nothing, stumbling on blindly through the dripping woods until the first light of morning.

But by far the worst of the night's horrors was the fire at the bridge. Minnie Chambers, the girl who clung to the roof of the Cambria works, said later that she could hear screaming from the bridge all through the night. William Tice, who owned a drugstore on Portage Street, described what he saw soon after he had been fished out of the water near the bridge.

"I went up on the embankment and looked across the bridge, which was filled full of debris, and on it were thousands of men, women, and children, who were screaming and yelling for help, as at this time the debris was on fire, and after each crash there was a moment of solemn silence, and those voices would again be heard crying in vain for the help that came not. At each crash hundreds were forced under and slain.

"I saw hundreds of them as the flames approached throw up their hands and fall backward into the fire, and those who had escaped drowning were reserved for the more horrible fate of being burned to death. At last I could endure it no longer, and had to leave, as I could see no more."

Frank McDonald, a railroad conductor who apparently kept on watching, said, "They reminded me of a lot of flies on flypaper, struggling to get away with no hope and no chance to save them."

Actually, for anyone to see much of what was going on was extremely difficult, with the rain pouring down, the dark, the smoke, and the wild flames. One after another, houses had been swept against the pileup and quickly took fire. Out of them, crawling on hands and knees, climbing, jumping from place to place, helping one another, small, dark figures had appeared, now silhouetted sharply against the high, wind-whipped flames, now invisible against the black shadow of mangled debris, now emerging again from the smoke and groping their way toward the ends of the bridge. That there looked to be thousands of them and that they seemed insectlike is understandable enough; but evidence is that, at most, perhaps 500 to 600 people were driven into the burning heap, and though exact figures were never settled on, it is likely that all but about 80 of them managed to escape.

A good many escapes were made thanks to the courage of bystanders who rushed in to help. They lifted old people and children from the windows of half-shattered houses. They helped carry the badly injured across the wreckage to the hillsides.

A girl named Rose Clark was trapped near one end of the bridge, half submerged under water, with a broken arm and a broken leg which was pinned down by timbers. A group of men had worked for several hours to free her leg but without success and the fire kept spreading closer. For a short while there was talk among them of cutting her leg off, rather than letting her burn to death, and for a few tense minutes, when the flame was almost on top of them, it looked as though they would have to. But the leg came free at last, and they carried her to safety.

The fire burned on through the night, and would be still blazing when morning came. In little towns miles away downriver and on the other side of the mountains, people could see a strange, shimmering, blood-red glow in the sky.

But even for those who had somehow succeeded in getting to the high ground in time, even for those who were uninjured or were lucky enough to have a roof to sleep under, there was the indescribable agony of remembering what they had seen, and not knowing what had become of others. No one really knew for sure the extent of what had happened, but they knew it had been terrible beyond belief, and if the whereabouts of someone was not known, then only the worst could be imagined. All that could be done now was to wait for morning, and hope.

VI

A message from Mr. Pitcairn

Mr. Robert Pitcairn's private car had been attached to east-bound passenger train Number 18 shortly before noon that morning and rolled out of Pittsburgh's Union Station about an hour later. Mr. Pitcairn was on his way to Lilly to see how serious the storm damage was there, and to look things over at Johnstown and South Fork on the way.

Messages about trouble along the line had been coming in to his office since early morning, including one about the dam. Pitcairn had read it and thought little more of it. First of all, he could not quite understand how Colonel Unger could be sending such warnings, since he knew perfectly well that Colonel Unger had no telegraph wire at the club and that the telephone line was not open yet. And secondly, as he would say later, he simply "paid little attention to any reports about the South Fork dam, as they had been made perhaps nearly every year."

When later messages came in from South Fork, from agent Dougherty and Pitcairn's old friend J. P. Wilson, Pitcairn was already on his way east.

Pitcairn's knowledge of the dam went back more than thirty

174

years, to the time when the Pennsylvania had first bought it. His old boyhood friend Andy Carnegie had gotten him a job on the railroad, as a ticket agent at Cresson, not long before that. He and Carnegie had been telegraph operators together in Pittsburgh; then they went with the Pennsylvania. Later on, when Andy quit his job as head of the Pittsburgh Division to go into business for himself, Pitcairn had been named to replace him.

But his first real interest in the dam began when it broke in 1862 and wrecked a lot of railroad property in South Fork. Then nearly twenty years later, when the South Fork club finished its restoration and there was talk in the valley about leaks at the base of the dam, Pitcairn had gone up to see for himself, taking along several of his own people from South Fork. They had given the dam what he felt was a thorough enough going-over. Benjamin Ruff had walked with them, saying that what everyone called leaks were actually springs that came from near the ends of the dam. Ruff also promised that he would strengthen things some, and then they all shook hands and went home.

"The only point we were afraid of," Pitcairn said later, "was the leaks at the bottom of the dam increasing." And he was evidently afraid (or cautious) enough to ask Wilson and others in the area to keep an eye out for him. Whether his subsequent membership in the South Fork Fishing and Hunting Club was, even in part, his way of keeping his own eye out, is not known. Though it seems highly doubtful, since, unlike Daniel J. Morrell, he had every sort of social and business reason for wishing to be there.

Now his train headed out of Pittsburgh along the muddy Monongahela, past the towering black Edgar Thomson works at Braddock, where the British general by the same name had suffered his famous defeat; then on through one little town after another, East McKeesport, Irwin, Jeannette, Greensburg, and out into open hill country. He had been over the route maybe a thousand times during his years with the railroad. He knew every bend, bridge, siding, every water tower, coal tipple, every depot, every barn and farmhouse along the horizon. He could not claim to know all the men, that would be impossible, but he knew the good part of them, and certainly every last one of them knew him. He was their supreme commander. His word was law from Altoona to Pittsburgh, and the portly frame, the bullet head, the pince-nez glasses and walrus mustache were far better-known among them than the rather inconspicuous features of the man who was then President of the United States. And if it were a question as to which one wielded the most authority, there would have been some debate.

Men who had been with the Pennsylvania for a dozen years or more still talked about Pitcairn during the "Great Strike" of '77. A lot of them felt that he had been the chief cause for what happened in Pittsburgh. The strike had begun on the B & O in Baltimore and had been spreading fast. Times were hard, and wages had been cut. But Pitcairn had chosen that particular moment to institute new practices on the division which would have meant increased work loads and even more layoffs. "The men were always complaining about something," he would say later. When it was all over, a pitched battle had been fought between militia and a mob of strikers and unemployed, a good part of downtown Pittsburgh had been burned to the ground, and fifty-seven people had been killed. The men had gone back to work, having gained nothing. And Robert Pitcairn's hold on them was stronger than ever.

He liked to say that the railroads (by which he really meant his railroad) were "the heart, blood, veins, and arteries of Pittsburgh," which, of course, put him in a most important position indeed.

Pitcairn was in his mid-fifties. He was an elder of the fashionable new Shady Side Presbyterian Church and a man of considerable financial means. For aside from his earnings on the railroad, he had also managed to put together a good-sized fortune on the side, largely by backing his inventive friend George Westinghouse. Slabtown and the squalid back streets where he and Carnegie and Phipps had grown up were far, far behind him now.

Sitting in the upholstered splendor of his private car, he looked appropriately substantial, and quite tired. He had been up late the night before at the telegraph instrument in his home making inquiries about the weather between Pittsburgh and Altoona. Then he had gone off to town earlier than usual that morning, and with the news of the storm growing worse every hour, it had been a difficult day ever since. But now, watching the landscape sweep past his window, he began to realize just how serious things were.

The rain was coming down in wild, silvery sheets. The whole countryside was awash. Hillsides were mortally veined with angry little creeks. Fields were covered with water that looked to be a foot deep or more. At Latrobe, at the foot of Chestnut Ridge, the Loyalhanna was twice its normal size and well over its banks. Ten miles farther, he saw the Conemaugh for the first time and knew he was up against something unlike anything he had ever experienced. His train was moving very slowly by this time, following the course of the river where it cut through the ridge. On the other side of the ridge, at the village of Bolivar, people were out along the riverbanks watching the torrent rush by. At New Florence the water had spilled through the lowlands, flooding miles of woods and meadows. Pitcairn thought he could actually see the water rising, it was coming up so fast.

After New Florence the train pulled through little Nineveh, where men and boys in gum coats, their collars turned up against the biting wind, their hats dripping with rain, stood beside the track watching the cars clack by.

Then the train started into the breath-takingly beautiful Conemaugh Gap, or Packsaddle, the one pass through Laurel Hill to Johnstown. The railroad ran well above the river here, even with the river in its present condition, but above the tracks the mountainside loomed up another 1,500 feet.

Several miles farther on in the gorge, at a place called Sang Hollow, about four miles from Johnstown, they stopped. The time by now was about five after four.

Pitcairn climbed down from his car and went up to the tower to find out what the trouble was. The operator told him the lines east had gone dead; they tried again several times to get Johnstown, but it was no use. The operator said he could not let them through without clearance, which, according to the rules, was exactly what he was meant to say, even to Pitcairn.

"I was about making up my mind to proceed cautiously, running carefully, to find the trouble," Pitcairn said later, "when looking east, I saw some debris. The water before this had been muddy, but very little drift. The debris attracted my attention from its singular appearance, being broken up wood entirely, and in very small pieces. In a short time, the telegraph poles commenced to break down, and threatened to take the tower down with it."

Then they saw a man coming down the river on some debris, moving very fast. Pitcairn thought the water must have been going by at about fifteen miles an hour. They saw more people coming, hanging on to telegraph poles or what appeared to be parts of buildings or just being swept along and trying desperately and futilely to swim. Pitcairn and the others rushed out to do what they could to save them, but the river carried them off and out of sight.

"I returned to the telegraph office to see what word I could get, when the people came down by the scores; the water rising very rapidly, and men, women, and children on the drift, and we perfectly helpless."

By this time most of the male passengers on board the train were out on the riverbanks doing everything possible to help. They got hold of long poles and big limbs and held them out over the current as far as they could, hoping maybe the people going by could reach them. They threw ropes, and at one point, one of them actually stripped off his coat and jumped into the water to save a mother and her small child.

He was a rugged, seventeen-year-old Pittsburgh boy named Bill Heppenstall, who was on his way back to school at Bellefonte, Pennsylvania, after being home ill. A small house had lodged momentarily in some overhanging trees. The men heard a baby crying, but the house was too far out to reach. Heppenstall decided he would go in and get the child. The others tried to talk him out of it, but he got the bell cord out of one of the cars, tied it around him, and swam out to the house. In no time he was back with the child. There were great cheers from the crowd. But he then told them the mother was still back there and started into the water again, this time taking a railroad tie along with him to help hold her up. Just as he got her to shore the house tore loose from the trees and went spinning off downstream.

By the time it began getting dark, the operator at Sang Hollow had counted 119 people going by, dead and alive. Despite everything they had tried, the men on the riverbank had been able to rescue only seven.

About six o'clock Pitcairn ordered the train back down to New Florence. The water was still high, but it did not seem to be getting any higher. He had decided to take the passengers back to Pittsburgh, giving them the option to stop off at New Florence if there were any accommodations to be had.

But before leaving, Pitcairn got off a message to Pittsburgh. It was directed to the editors of the morning papers, and its exact wording remains unclear. But sometime between five thirty and six the news was out and on the wire. A dam had failed at South Fork and caused a disastrous flood at Johnstown. By then Pitcairn had faced up to the awful realization of what had gone wrong.

His train rolled ever so slowly back through the gorge, reaching New Florence by perhaps six thirty. The first thing he did there was to write out a still longer and more detailed message, which he then put aside, in the hope that some further word might come in from Johnstown itself. So for the next several hours they sat and waited. The rain hammered down outside; men kept coming in and out talking of more bodies found or the few half-drowned souls they had been able to drag ashore.

The village sat well back from the river, on high, dry ground. Only a few houses near the river were under water, and few citizens were suffering any serious discomforts. As a result the streets were filled with people going to and from the river, standing in doorways, talking to passengers from Pitcairn's train, or gathered in groups looking at the dull, red glow in the sky to the east.`

About ten o'clock Pitcairn received word from Johnstown by way of Sang Hollow. One of his men in Johnstown, a W. N. Hays, had managed to get from Johnstown to Sang Hollow on foot. Apparently he had been on the hillside above the west end of the bridge and was able to make his way down the tracks above the rampaging river. Once he reached Sang Hollow, the message was put on the wire to New Florence.

Pitcairn was told how things were at Johnstown, and he then sent a second message to Pittsburgh, which would be quoted in the papers there at some length. He reported the number of bodies that had been counted going by at Sang Hollow. He said there was no way clear to Johnstown, but that his information was that the city was "literally wiped out." He said that the debris at the stone bridge was reported to be forty feet high and that it was burning.

Then he said, "I fear there will be terrible suffering among those saved which should be relieved as soon as possible. In the interest of humanity I think a public meeting should be called early tomorrow to send food, clothing, etc. to those poor people which we will be glad to forward to Johnstown . . . as soon as we can get a clear track there."

This message, like the one before it, went right on the wire. Before midnight the story was across the country:

> Pittsburgh, Penn. May 31—A rumor, loaded with horror, holds this city in dreadful expectancy tonight. It is said that the bursting of a reservoir, just above Johnstown, a flourishing place in Cambria County, had flooded the town and swept at least 200 of her citizens to death. The news is of a very uncertain character, there being no communication with the district were the flood is reported to have occurred, all the wires being down . . . There is no way to get to the scene of the disaster and full particulars are not expected tonight.

But the fact was that the rush to Johnstown had already begun hours before. Two trains had been chartered by five Pittsburgh newspapers, and the first of them, the one with the *Dispatch* and *Times* men on board, started out from Union Station a few minutes after seven. The second, chartered by the *Post*, the *Commercial-Gazette*, and the *Chronicle-Telegraph*, followed almost immediately after. In New York, Philadelphia, Boston, in Chicago, Cleveland, and St. Louis, reporters picked up their hats and coats and went directly to the nearest depot, taking no time to pack or anything else. Some of them were still in evening clothes after a night at the theater.

The trains from Pittsburgh got no farther than Bolivar, where the men piled out into the rain and moved among the crowds gathered at the station and along the dark edge of the river. They picked up stories of the bodies and wreckage that had been washing past, about the few rescues that had been made, and the horrid things people had seen happen.

Not long before dark a man and two women had been seen rounding the bend upstream. They were on a raft of some sort, a barn roof most people thought it was, and they were coming on fast, the women down on their knees, the man with his arms around them and looking about for something to grab on to. That had been before the bridges went, and the men on the bridges had been hanging ropes down for the people in the river to get hold of. When the raft shot by under the first bridge, the man reached out for the rope but missed. Then he and the two women were heading for the second bridge, and everyone along the shore line was rooting for them as they watched him telling the women to try for the next rope. As they came under the second bridge, he made a lunge for the rope, got it, and was jerked violently off balance; but seeing that the women had missed, he let go and fell back down on the raft again. The current then swept them toward the bank, where he was able to catch hold of a tree. With an immense effort he managed to pull the two women into the tree with him, but at almost the same instant a large section of the bridge upstream let go with a sudden crash. It came careening down the river, smashed into the tree, carried it away, and drowned the man and the two women.

Everyone in Bolivar had seen the whole thing and they wanted to tell exactly how it happened. Some people in the crowd said they knew the man and said his name was Young. Others said they thought the women looked like mother and daughter, and that they could be heard praying as they went by. The newspapermen wrote it all down, asking questions, taking names.

It was too dark to see much by the river, but the rush of the water could still be heard plain enough, and tiny, dim specks of light could be seen moving through the trees along the shore where men with lanterns were still watching for possible signs of life.

Johnstown was still twenty miles away. Among the newspapermen there was talk about what to do next. The tracks from Bolivar on were under water and not safe enough to take the train any farther. Most of the men decided to push on in the direction of New Florence, some by foot and some in wagons. The ride up from Pittsburgh had taken quite a long while, with conditions what they were. It was ten thirty when they had pulled into Bolivar. By the time they had slogged through the rain and dark to New Florence, it was getting on toward three in the morning.

Mud-spattered, dead-tired, cold, wringing-wet, they moved into whatever dry space there was left in the little town and began interviewing everyone who was willing to talk, which was just about everyone. Several of them got hold of wires to Pittsburgh and started filing their stories.

At that point about all they could say was that every sign was that "hundreds if not thousands" of people had been killed in "an appalling catastrophe." They reported rumors of panic-stricken people fleeing through the woods from the scene of the disaster and of the number of people who had been seen going by in the river at New Florence (counts varied, but eighty-five seems about average). And they sent back what information they could pick up concerning the dam, a good deal of which was inaccurate. Several reporters had the dam 110 feet high and the lake as much as eight miles long and three miles wide. But they did have the name of the South Fork Fishing and Hunting Club by then, reported it accurately, and added that the Pennsylvania Railroad's engineers had inspected the dam once a month, which suggests that Pitcairn was also, by then, doing some talking himself. "Investigations showed that nothing less than some convulsion of nature would tear the the barrier away and loosen the weapon of death," one reporter put on the wire.

About four o'clock there was great excitement when a man from Johnstown, a carpenter named McCartney, and his wife came staggering out of the night. He said they had left Johnstown right after the flood struck and had been walking ever since. He told them there was hardly a building left standing in Johnstown, and, in general, substantiated the wildest of the rumors that had been circulating since the night began.

Sometime soon after four Pitcairn decided that there was no use staying in New Florence any longer. He boarded his private car, and with the passengers back on board again, his train started for Pittsburgh. It was also about the same time that several of the newspapermen decided that if the McCartneys could make it across the mountain on foot, they could too, and they set off through the woods. With luck, they figured, they could be on the hill above the city by daybreak.

VII

In the valley of death

It was nearly morning when the strange quiet began. Until then there had been almost no letup from the hideous sounds from below. Few people had been able to sleep, and several of the war veterans were saying it was the worst night they had ever been through.

But in the last chill hour before light, the valley seemed to hang suspended in an unearthly stillness, almost as unnerving in its way as everything else that had happened. And it was then, for the first time, that people began to realize that all those harsh, incessant noises which had been such a part of their lives—mill whistles screeching, wagons clattering over cobblestones, coal trains rumbling past day and night—had stopped, absolutely, every one of them.

About five the first dim shapes began emerging from the darkness. But even by six very little stood out in detail. There were no shadows, no clear edges to anything. Some survivors, years later, would swear it had been a bright, warm morning, with a spotless blue sky, which, after the night they had been through, it may well

have seemed. But the fact is that though the rain had at last
stopped, the weather on that morning of June 1 was nearly as foul
as it had been the morning before. The valley looked smothered in
a smoky gray film. The hills appeared to be made of some kind of
soft, gray-green stuff and were just barely distinguishable from a
damp, low sky that was the color of pewter. Odd patches of the
valley were completely lost in low-hanging ribbons of mist, and the
over-all visibility was reduced to perhaps a mile at best.

Still the view that morning would be etched sharply in the
memories of everyone who took it in. Along the Frankstown Road
on Green Hill some 3,000 people had gathered. On the rim of Prospect Hill and on the slopes above Kernville, Woodvale, and Cambria City the crowds were nearly as big. Chilled to the bone, hungry, many of them badly injured, hundreds without shoes or only
partly clothed against the biting air, they huddled under dripping
trees or stood along narrow footpaths ankle-deep in mud, straining
their eyes to see and trying hard to understand.

Spread out below them was a vast sea of muck and rubble and
filthy water. Nearly all of Johnstown had been destroyed. That it
was even the same place was very difficult to comprehend.

There were still a few buildings standing where they had been.
The Methodist Church and the B & O station, the schoolhouse on
Adams Street, Alma Hall, and the Union Street School could be
seen plain enough, right where they were meant to be. The Iron
Company's red-brick offices were still standing, as was Wood,
Morrell & Company next door. But everywhere else there seemed
nothing but bewildering desolation. The only immediately familiar
parts of the landscape were the two rivers churning toward the
stone bridge, both still swollen and full of debris.

From Woodvale to the bridge ran an unbroken swath of
destruction that was a quarter of a mile wide in places and a good
two miles long. From Locust Street over to the Little Conemaugh
was open space now, an empty tract of mud, rock, and scattered
wreckage, where before saloons, stores, hotels, and houses had been
as thick as it had been possible to build them. Washington Street
was gone except for the B & O station. Along Main, where a cluster
of buildings and gutted houses still stood like a small, ravaged island, the wreckage was piled as high as the roofs of the houses.

At the eastern end of town all that remained between Jackson and Clinton was a piece of St. John's Convent. At the corner of Jackson and Locust the blackened rafters of St. John's Church were still smoking, and where the Quinn house had stood, fifty feet away, there was now only a jumble of rubbish.

Across the Stony Creek, Kernville had been swept clean for blocks. Virtually everything was gone, as though that whole section had been hosed down to the raw earth. The entire western end of town, near the Point, was now a broad, flooded wasteland. Every bridge was gone except the stone bridge and against it now lay a good part of what had been Johnstown in a gigantic blazing heap.

Below the stone bridge the ironworks, though still standing, looked all askew, with stacks toppled over and one of the biggest buildings caved in at the end as though it had been tramped on by an immense heel. Cambria City had been ravaged past recognition. At least two-thirds of the houses had been wiped out, and down the entire length of its main street a tremendous pile of mud and rock had been dumped.

With the light the first small groups of people who had survived the night down in the drowned city could be seen making their way across the debris, most of them heading for Green Hill, where dry ground could be reached without having to cross a river. From Alma Hall and the Union Street School they came in steady little batches, moving up and down over the incredible flotsam. Then, at the same time from Frankstown Road and other near hills, men started moving down into town. And as they came closer, the dim sweep of destruction began to take on a different look. Slowly things came into ever sharper focus.

The Morrell house could be seen with part of its side sheared off. Dr. Lowman's house stood alone on the park, the only big house still there, but its two-story front porch had been squashed and every window punched in. Colonel Linton's place on lower Main looked as though it had been blasted in two by dynamite, and the black span of an iron bridge was resting where the yard had been. Beyond, houses were dumped every which way, crushed, broken, split clean in half, or lying belly up in the mire, with their floor beams showing like the ribs of butchered animals.

Telephone poles, giant chunks of machinery, trees with all their bark shredded off, dead horses and pieces of dead horses, and countless human corpses were strewn everywhere. "Hands of the dead stuck out of the ruins. Dead everywhere you went, their arms stretched above their heads almost without exception—the last instinct of expiring humanity grasping at a straw," wrote George Gibbs, one of the reporters from the *Tribune*.

And now, too, all the litter of thousands of lives could be seen in sharp detail. Shattered tables and chairs, tools, toys, account books, broken dishes, chamber pots and bicycle wheels, nail kegs, bedquilts, millions of planks and shingles were thrown up in grotesque heaps ten, twenty, thirty feet high, or lay gently shifting back and forth in huge pools of water that covered much of the valley floor like a brown soup.

"It were vain to undertake to tell the world how or what we felt, when shoeless, hatless, and many of us almost naked, some bruised and broken, we stood there and looked upon that scene of death and desolation," David Beale wrote.

The flood and the night that had followed, for all their terror and destruction and suffering, had had a certain terrible majesty. Many people had thought it was Judgment Day, God's time of anger come at last, the Day of Reckoning. They thought that the whole world was being destroyed and not just Johnstown. It had been the "horrible tempest," with flood and fire "come as a destruction from the Almighty." It had been awful, but it had been God Awful.

This that lay before them now in the dismal cold was just ugly and sordid and heartbreaking; and already it was beginning to smell.

Rescue parties got to work bringing the marooned down from rooftops and went searching among the wreckage for signs of life. Men scrambled over piles of debris to get to the upstairs windows of buildings that looked as though they might fall in at any minute. They crawled across slippery, cockeyed roofs to squeeze through attic windows or groped their way down dripping back hallways where the mud was over their boot tops. It was treacherous work and slow going. Walls were still falling in and fires were breaking out.

At the stone bridge, gangs of men and boys, many of whom had been there through the night, were still working to free people trapped alive within the burning pile. Young Victor Heiser, who had succeeded in reaching solid ground after his night in a Kernville attic, had made his way down the west bank of the Stony Creek as far as the bridge, where, as he wrote later, "I joined the rescue squads, and we struggled for hours trying to release them from this funeral pyre, but our efforts were tragically hampered by the lack of axes and other tools. We could not save them all. It was horrible to watch helplessly while people, many of whom I actually knew, were being devoured in the holocaust."

Across the whole of the valley the dead were being found in increasing numbers. And as the morning passed, more and more people came down from the hillsides to look at the bodies, to search for missing husbands and children, or just to get their bearings, if possible. They slogged through the mud, asking after a six-year-old boy "about so high," or a wife or a father. They picked their way through mountains of rubbish, trying to find a recognizable landmark to tell them where their house or store had been, or even a suggestion of the street where they had lived. Or they stood silently staring about, a numb, blank look on their faces. Over and over, later, when the day had passed, people would talk about how expressionless everyone had looked and how there had been so few people crying.

There was some shouting back and forth among the men. People who had been separated during the night would suddenly find one another. "What strange meetings there were," wrote one man. "People who had hardly known each other before the flood embraced one another, while those who found relations rushed into each other's arms and cried for very gladness that they were alive. All ordinary rules of decorum and differences of religion, politics and position were forgotten."

Lone stragglers went poking about looking for only they knew what, many of them strangely clad in whatever odd bits of clothing they had been able to lay hands on. One man, hatless and with a woman's red shawl across his shoulders, came limping along in his stocking feet, using a piece of lath for a cane. He was looking for his wife, Mrs. Brinker, who, as he would soon discover, had survived the night inside the Methodist parsonage and who had long since given him up for dead.

People recovered some pathetic belonging or other and carried it carefully back to high ground or began building little personal piles of salvage. There was no order to what went on, no organization, and not much sense. Most people were unable even to look after themselves; they were stunned, confused, trying, as much as anything, to grasp what had happened and what was left of their lives. Where they went from there was something they were not yet ready to think about. Many of them struck off into the country, with no special destination. They just kept walking for hours, looking for food or a dry place to lie down for the night, or, very often, just trying to put as many miles as possible between themselves and the devastated city. They were afraid of the place and wanted no more part of it.

The problems to be faced immediately were enormous and critical. People were ravenously hungry, most everyone having gone twenty-four hours or more without anything to eat, and now there was virtually no food anywhere. The few provisions uncovered among the ruins were nearly all unfit for eating, and what little else people had was given to the injured and to the children. Moreover, there was no water that anyone felt was safe to drink. Thousands were homeless, hundreds were severely injured. Mrs. John Geis, for example, little Gertrude Quinn's grandmother, had had her scalp torn off from her forehead back to the nape of her neck. Hundreds of others were dazed by lack of sleep or in a state of shock. Dozens of people, as a result of exposure, were already in the early stages of pneumonia. There was almost no dry clothing to be had and no medicines.

People had no money, except what change they may have had in their pockets at the time the water struck, and even if they did, there were no stores left at which to buy anything. There was no gas or electric light. Fires were burning in a dozen different places, and no one knew when a gas main might explode. Every telegraph and telephone line to the outside world was down. Bridges were gone, roads impassable. The railroad had been destroyed. And with the dead lying about everywhere, plus hundreds of carcasses of drowned horses, cows, pigs, dogs, cats, birds, rats, the threat of a violent epidemic was very serious indeed.

But by noon things had begun to happen, if only in a small way. Rafts had been built to cross the rivers and to get over to those buildings still surrounded by water. People on the hillsides whose houses had escaped harm and farmers from miles out in the country began coming into town bringing food, water, and clothing. At the corner of Adams and Main milk was passed out in big tinfuls. Unclaimed children were looked after. A rope bridge had been strung across the Little Conemaugh near the depot, and, most important of all as it would turn out, up at the Haws Cement Works, on the hill at the western end of the stone bridge, several bedraggled-looking newspaper correspondents had established headquarters in a coal shed and were in the process of rigging their own wire down the river to Sang Hollow.

The men had reached Johnstown about seven in the morning, and like everyone else were cold, dirty, hollow-eyed from no sleep. There remains some question as to which of them arrived first, but William Connelly, who was the Associated Press correspondent in western Pennsylvania, Harry Orr, a telegraph operator for the A.P., and Claude Wetmore, a free-lance reporter working for the New York *World*, are generally given the credit. Others kept straggling in from New Florence through the rest of the day. But until nightfall the major stories were still being filed out of the little railroad crossing on the other side of Laurel Hill.

> New Florence, Pa. June 1 . . . Seven bodies have been found on the shore near this town, two being on a tree where the tide had carried them. The country people are coming into the news centers in large numbers, telling stories of disaster along the river banks in sequestered places . . . The body of another woman has just been discovered in the river here. Only her foot was above the water. A rope was fastened about it and tied to a tree . . . R. B. Rogers, Justice of the Peace at Nineveh, has wired the Coroner at Greensburg that 100 bodies have been found at that place, and he asks what to do with them.

That afternoon, at three, a meeting was called in Johnstown to decide what ought to be done there. Every able-bodied man who could be rounded up crowded into the Adams Street schoolhouse. The first step, it was quickly agreed, was to elect a "dictator." John Fulton was the obvious choice, but he was nowhere to be found, so it was assumed he was dead, which he was not. He had left town some days earlier and was at that moment, like hundreds of others, trying desperately to get to Johnstown.

The second choice was Arthur J. Moxham, a remarkable young Welshman who had moved to Johnstown a few years before to start a new business making steel rails for trolley-car lines. In the short time he had been there Moxham had about convinced everyone that he was the best newcomer to arrive in the valley since D. J. Morrell. His business had prospered rapidly, and it was earlier that spring that he had opened a sprawling new complex of mills up the Stony Creek beside the new town he had developed. He named the business the Johnson Steel Street Rail Company, after his lively young partner, Tom L. Johnson, who, in turn, had named the town Moxham. They paid their men regularly each week, in cash, and did not maintain a company store—all of which had had a marked impact on the town's economic well-being and a good deal to do with their own popularity.

Both men were energetic, able executives. Both were already wealthy, and both, interestingly enough, were devout followers of the great economic reformer of the time, Henry George, and were equally well known in Johnstown for their impassioned oratory on George's single-tax scheme.

Moxham was a fortunate choice. He took charge immediately and organized citizens' committees to look after the most pressing and obvious problems. Morgues were to be established under the direction of the Reverends Beale and Chapman. Charles Zimmerman and Tom Johnson were put in charge of removing dead animals and wreckage. (That anyone could have even considered cleaning up the mess at that point is extraordinary, but apparently the work began right away, against all odds, against all reason. Trying to bail the rivers dry with buckets would have seemed not much more futile.)

Dr. Lowman and Dr. Matthews were responsible for establishing temporary hospitals. Captain Hart was to organize a police force. There was a committee for supplies and one for finance, to which George Swank and Cyrus Elder were assigned.

Captain Hart deputized some seventy-five men, most of whom were employees of the Johnson Company sent down from Moxham. They cut tin stars from tomato cans found in the wreckage, threw a cordon around the First National and Dibert banks, and, according to a report made days later, recovered some $6,000 in cash from trunks, valises, and bureau drawers lying about.

As dusk gathered, the search for the living as well as the dead went on in earnest. There seemed to be no one who was not missing some member of his family. James Quinn had already found little Gertrude, but he was still looking for his son Vincent, his sister-in-law and her infant son, and Libby Hipp, the nursegirl, though he had little hope of finding any of them except his son. That Gertrude was alive seemed almost beyond belief.

He and his other daughters had been luckier than most and had spent the night in a house on Green Hill. At daybreak he had been outside washing his face in a basin when his sister, Barbara Foster, came running up shouting that she had found Gertrude. She had seen her on the porch of the Metz house, still speechless with fright, still unidentified, and almost unrecognizable with her blonde hair tangled and matted with mud, her dark eyes quick with terror. Quinn at first found it impossible to accept what he heard, but started off at a run, the lather still on his face, and the other little girls running behind.

"When he came near the house," Gertrude wrote later, "I saw him and recognized him at once. I fairly flew down the steps. Just as he put his foot on the first step, I landed on his knee and put both my arms around his neck while he embraced me."

Quinn gathered up the child. They both began crying. A small crowd had assembled by now, on the porch and on the street below, and the scene caused several people to break down for perhaps the first time. Then there was a lot of handshaking and Quinn set off with his children to find his son.

Victor Heiser had spent most of the day searching for his mother and father, hoping against hope that somehow they had come through it all alive and in one piece. His own survival seemed such a miracle to him that he could not help feeling there was a chance they might be somewhere in the oncoming darkness looking for him.

At the bridge late in the afternoon an old man and his daughter were rescued from a house wedged among the burning wreckage. The old man made quite a reputation for himself when, on being helped down into a rowboat, he asked his rescuers, "Which one of you gentlemen would be good enough to give me a chew of tobacco?" And on the hillside a few hundred yards away two young ladies who had been stripped naked by the flood were found cowering in the bushes, where they had been hiding through the long day, too ashamed to venture out before dark.

Cyrus Elder's wife and daughter were missing. Horace Rose did not learn until late in the afternoon that the two sons, Winter and Percy, from whom he had been separated during the flood were still alive, and that he was the only member of his large family who had even been injured.

His neighbor John Dibert had already been identified among the dead, as had Mrs. Fronheiser, whom Rose had last seen in her window next door. The bodies of Samuel Eldridge, one of the best-known policemen in town, and Elizabeth Bryan of Philadelphia, who had been on the *Day Express*, had also been found. But of the other dead found only a small number had as yet been identified for sure.

At the Adams Street schoolhouse and a saloon in Morrellville, where the first two emergency morgues had been opened, the bodies were piling up faster than they could be properly handled. They came in on planks, doors, anything that would serve as a stretcher, and with no wagons or horses as yet on hand, the work of carrying them through the mud and water was terribly difficult.

Each body was cleaned up as much as was possible, and any valuables found were put aside for safekeeping. Those in charge tried hard to maintain order, but people kept pushing in and out to look, and the confusion was terrific.

"We had no record books," David Beale wrote, "not even paper, on which to make our records, and had to use with great economy that which we gathered amid the debris or happened to have in our pockets."

One way or other the bodies were numbered and identified, whenever that was possible. Many were in ghastly condition, stripped of their clothes, badly cut, limbs torn off, battered, bloated, some already turning black. Others looked as though they had suffered hardly at all and, except for their wet, filthy clothes, appeared very much at peace.

A Harrisburg newspaperman named J. J. MacLaurin, who had been near Johnstown at the time the flood struck, described a visit to the Adams Street School early Saturday afternoon, where he counted fifty-three bodies stretched on boards along the tops of the desks. "Next to the entrance lay, in her damp clothing, the waiter-girl who had served my last dinner at the Hulbert House, with an-other of the dining room girls by her side."

How many dead there were in all no one had any way of knowing, since there was, as yet, almost no communication be-tween various parts of town. But wild estimates were everywhere by nightfall, and with more bodies being discovered wherever the wreckage had been pulled apart, it was generally agreed that the final count would run far into the thousands. Some were saying it would be as much as 10,000 by the time the losses were added up from South Fork to Johnstown, and few people found that at all hard to believe. What may have happened on down the river at Nineveh or New Florence or Bolivar was anyone's guess.

Within another day the Pennsylvania station and the Presby-terian Church, a soap factory, a house in Kernville, the Millville School, and the Catholic Church in Cambria City would be con-verted into emergency morgues. But it would be a week before things got down to a system at these places, and not for months would there be a realistic count of the dead. Actually, there never would be an exact, final count, though it is certain that well over 2,000 people were killed, and 2,209 is generally accepted as the offi-cial total.

Hundreds of people who were lost would never be found. One out of every three bodies that was found would never be identified beyond what was put down in the morgue records. With all the anguish and turmoil of the first few days, such entries were at best a line or two.

...11. Unknown.
        A female. "FL.F." on envelope.
...17. Unknown.
        A man about fifty years of age. Short hair, smooth
        face.
...25. Unknown.
        Female. Light hair. About fifteen years.

Later, more care would be taken to be as explicit as possible.

...181. Unknown.
        Female. Age forty-five. Height 5 feet 6 inches.

Weight 100. White. Very long black hair, mixed with grey. White handkerchief with red border. Black striped waist. Black dress. Plain gold ring on third finger of left hand. Red flannel underwear. Black stockings. Five pennies in purse. Bunch of keys.

...182. Unknown.

Male. Age five years. Sandy hair. Checkered waist. Ribbed knee pants. Red undershirt. Black stockings darned in both heels.

. .204. Unknown.

Male. Age fifty. Weight 160. Height 5 feet 9 inches. Sandy hair. Plain ring on third finger of left hand (with initials inside "C.R. 1869.") Pair blood stone cuff-buttons. Black alpaca coat. Navy blue vest and pants. Congress gaiters. Red stockings. Pocketbook. Knife and pencil. $13.30 in change. Open-faced silver watch. Heavy plaited chain and locket. Inside of locket a star with S.H., words trade-mark alone a star. Chain trinket with Washington head. Reverse the Lord's prayer. Odd Fellow's badge on pin.

In all, 663 bodies would be listed as unknown. A few were not identifiable because they had been decapitated. Close to a hundred had been burned beyond recognition, and some so badly that it was impossible even to tell what sex they had been. And many of the bodies found in late June or on into the summer and fall would be so decomposed as to be totally unrecognizable.

Part of the problem, too, was the fact that on the afternoon of May 31 Johnstown had had its usual share of strangers in town, nameless faces even when they had been alive, foreigners who had been living there only a short time, tramps, traveling men new to the territory, passengers on board any one of the several trains stalled along the line, countrypeople who had decided to stay over after Memorial Day. They made up a good part of the unknown dead, and doubtless many of them were among those who were never found at all.

Among the known dead were such very well-known figures as Dr. John Lee; Theodore Zimmerman, the lawyer; Squire Fisher, the Justice of the Peace, and his entire family; C. T. Schubert, editor of the German newspaper; and Ben Hoffman, the hackman, who, according to one account, "always got you to the depot in plenty of time" and whose voice was "as familiar as train whistle, iron works, or the clock bells." (Hoffman had gone upstairs to take a nap shortly before the flood struck and was found with his socks in his pockets.)

The Reverend Alonzo Diller, the new rector of St. Mark's Episcopal Church, was dead, along with his wife and child. George Wagoner, who was a dentist as well as a part-time preacher, and so one of the best-known men in town, was dead, as were his wife and three daughters. Emil Young, the jeweler, was dead; Sam Lenhart, the harness dealer, was dead; Henry Goldenberg, the clothier, Arthur Benshoff, the bookseller, Christian Kempel, the undertaker, were all dead. Mrs. Hirst, the librarian, lay crushed beneath a heap of bricks, slate, and books that stood where the public library had been.

Vincent Quinn was dead, as were Abbie Geis, her child, and Libby Hipp. Mrs. Cyrus Elder and her daughter Nan, Hettie Ogle and her daughter Minnie were dead, and their bodies would never be identified. George and Mathilde Heiser were dead.

Ninety-nine whole families had been wiped out. Three hundred and ninety-six children aged ten years or less had been killed. Ninety-eight children lost both parents. One hundred and twenty-four women were left widows; 198 men lost their wives.

One woman, Mrs. John Fenn, wife of the tinsmith on Locust Street, lost her husband and seven children. Christ Fitzharris, the saloonkeeper, his wife, father, and eight children were all drowned. Charles Murr and six of his children went down with his cigar store on Washington Street; only his wife and one child survived. In a house owned by John Ryan on Washington Street, twenty-one people drowned, including a man named Gottfried Hoffman, his wife and nine children.

At "Morgue A," the Adams Street schoolhouse, 301 bodies would be recorded in the logbooks. At the Presbyterian Church, which was "Morgue B," there would be 92; at "Morgue C," in the Millville schoolhouse, the total would come to 551 by the time the last entry was made ("Unknown") on December 3. And along with the prominent merchants and doctors, the lawyers and preachers, there were hundreds of people with names like Allison, Burns, Evans, Shumaker, Llewellyn, and Hesselbein, Berkebile, Mayhew, McHugh, Miller, Lambreski, Rosensteel, Brown, Smith, and Jones. They made up most of the lists, and in the town directory that was to have been published that June they were entered as schoolteacher, porter, or drayman, clerk, miner, molder, barber, sawyer, dressmaker, or domestic. Dozens of them were listed as steelworker, or simply as laborer, and quite often as widow.

In that part of the valley through which the flood had passed, the population on the afternoon of the 31st had been approximately 23,000 people, which means that the flood killed just about one person out of every ten. In Johnstown proper, it killed about one out of nine.

But there were no statistics for anyone to go on that Saturday night. It would be weeks before even a reasonably accurate estimate would be made on the death toll. The business of finding the dead just went very slowly. Young Vincent Quinn's body, for example, was not uncovered until June 7, buried beneath the wreckage in Jacob Zimmerman's yard. Victor Heiser's mother was found about the same time, her clothing still much intact, her body scarcely marked in any way; but the search for George Heiser went on for weeks after, and his body never was identified for certain. Toward the end of June a body was found which Victor was told was his father, but it was by then in such dreadful condition that he was not permitted to look at it.

In July there would be many days when ten to fifteen corpses would be uncovered. About thirty bodies would be found in August, including that of little Bessie Fronheiser; and so it would go on through the fall. In fact, for years to come bodies would keep turning up in and near the city. Two bodies would be found west of New Florence as late as 1906.

But by dark that Saturday only a small part of the dead had been accounted for, perhaps no more than 300 or 400, and only a very few had been buried. Most of the living found shelter well back from the city, on Prospect Hill or Green Hill, or on up the Stony Creek, where against the dark mountains tiny windows glowed like strings of orange lanterns. Or they walked to little towns like Brownstown, which was set in a high valley above Cambria City. Victor Heiser spent the next several nights there, along with more than 1,000 other refugees from the flood who were all housed, one way or other, by Brownstown's fifty-three resident families.

Houses, barns, stables, schools, churches, every remaining upright structure for miles around was put into service. Crude tents were fashioned from blankets and bedspreads. Lean-tos were built of planks and doors dragged from the wreckage.

One man later described smelling the odor of ham frying as he walked along the front street on Prospect Hill, and how he was invited into a small house "filled with a strangely composed company." There were two or three women who had been just recently rescued, and who were "pitiably pale, and with eyes ghastly at the flood horror." There was the hostess, who carried an infant on one hip, "a divine, a physician, a lawyer, two or three merchants," and several others. The dining room was too small to hold everyone, so they ate in shifts, waiting their turn out on the front porch. Below them, almost at their feet it seemed, lay the devastated valley.

The cold was nearly as cruel as it had been the night before. Pitch-blackness closed down over the mountainsides that crowded so close; but across the valley floor bonfires blazed, torches moved among the dark ruins, and the rivers and big pools of dead water were lighted by the fire that raged on at the stone bridge.

And with the deep night, for nearly everyone, came dreadful fear. There was the rational and quite justifiable fear of typhoid fever and of famine. It was entirely possible that a worse catastrophe than the flood itself could sweep the valley in a matter of days if help did not get through.

There were also rumors of thieves prowling through the night and of gangs of toughs who had come into the valley looking for trouble. Great quantities of whiskey were supposedly being found among the ruins, and drunken brawls were breaking out. People were warned to be on the lookout, that there would be looting and rape before the night was over; and men who had not slept since Thursday night took turns standing guard through the night, watching over their families or what little they may have had left of any earthly value.

Perhaps worst of all, however, was the wholly irrational fear of the very night itself and the nameless horrors it concealed. The valley was full of unburied dead; they were down there among the cold, vile remains of the city, waiting in the dark, and no one could get that idea out of his head for very long. If there were such a thing as ghosts, the night was full of them.

But despite it all, the hunger, the grief, the despair and fear, people gradually did what they had to; they slept. They put everything else out of their minds, for the moment, because they had to; and they slept.

Sunday the weather eased off. It was still cold, but the sky had cleared some, and for the first time in days it looked as though there would be no rain.

Sunday they began taking bodies across the Little Conemaugh in skiffs and carrying them to a plot on Prospect Hill where shallow graves were dug in the gravelly soil and the bodies buried without ceremony. (George Spangler, who had been night watchman at the First National Bank, wrote in his diary, "bisey holing the dead this day I hold 62 to the semitre.") Sunday a post office was set up at the corner of Adams and Main, and a clearinghouse where everyone who was still alive was meant to come in and register his name and tell what he knew about the rest of his family. Sunday the first patients were cared for in a temporary hospital on Bedford Street. And on Sunday the first relief trains got through.

A train from Somerset came in on the B & O tracks about daybreak. The other train, from Pittsburgh, had arrived at Sang Hollow about ten thirty Saturday night but had been unable to get any closer. From Sang Hollow to Johnstown there was practically nothing left of the old line. There were at least ten miles along the Pennsylvania where it was impossible to tell even where the tracks had been, and several of those ten miles could be accounted for between Sang Hollow and Johnstown.

The lonely little Sang Hollow depot had become the scene of great activity Saturday night, from about eleven on. Several boxcars had been unloaded and volunteers organized to start moving things upriver by hand. "The men carried the provisions on their backs," one participant wrote, "over landslides and the trackless roadbeds to points where handcars passed. All night long a procession of lights was moving to and fro from Sang Hollow to the stone bridge."

By morning nearly two carloads of supplies had been deposited at the western end of the bridge and work had begun on a rope bridge to get them over the Conemaugh. But more remarkable still was the fact that early Sunday, perhaps as early as eight in the morning, the Pittsburgh train itself came steaming up the valley clear to the stone bridge. So swiftly had the railroad swung back into action during the night that by dawn enough new track had been put down from Sang Hollow to start the train cautiously on its way. And as it crept through the ruins of Morrellville and Cambria City, men standing in the open doors of the boxcars passed out bundles of bread, cheese, and crackers to the ragged crowds that lined the tracks.

The supplies had left Pittsburgh about four Saturday afternoon. Pittsburgh had been in a frenzy since early that morning. The Allegheny River had risen sharply during the night, and the riverbanks and bridges were lined with people watching the wreckage sweep past. Already there were rumors that dead bodies had been fished out. "A sense of intense uneasiness pervaded the air," one man wrote.

There were still precious few facts to go on, but the papers were getting out a new edition every hour, and the news kept growing more and more alarming. Outside the newspaper offices, traffic was snarled by the crowds that pressed in to read the latest bulletins and kept calling for names.

At one o'clock a mass meeting was held at Pittsburgh's Old City Hall, at which Robert Pitcairn stood up and spoke briefly about what he had seen. "Gentlemen," he said in closing, "it is not tomorrow you want to act, but today; thousands of lives were lost in a moment, and the living need immediate help." Then there was a call for contributions. At the front of the hall two men using both hands took in $48,116.70 in fifty minutes. "There was no speech making," a reporter wrote, "no oratory but the eloquence of cash."

Wagons were sent through the city to collect food and clothing. Union Station looked like wartime, swarming with people and with train after train being loaded in the yards. The first train went out with twenty cars full. On board were some eighty volunteers of the "Pittsburgh Relief Committee," a dozen reporters, perhaps thirty police, and, according to one account, Mr. Durbin Horne, a member of the South Fork Fishing and Hunting Club, who was on his way to find out what had happened to several friends and relatives who had not been heard from since they left for the lake on Memorial Day.

When the rope bridge was finally finished Sunday morning at Johnstown, the men started over with their heavy loads, swaying precariously above the raging river. They came across one at a time and very slowly. And for the next several days, until the stone bridge was open again, they would keep on coming almost without stop.

Later on Sunday several good-sized boats would be hauled up the valley by train and put into service crossing the river, taking men and supplies over and bringing refugees back. On Sunday the boats ferried some 3,000 passengers, coming and going. Monday, they carried 7,000, along with supplies and dead bodies.

Wagons loaded down with salt pork, bedding, goods of every kind, rolled down flood-gullied roads from Ebensburg and Loretto, splashing up showers of gummy mud the color of a new baseball glove. Doctors and work crews started off from Altoona, where it was reported 5,000 people were milling about the railroad station. In dozens of little towns along the Pennsylvania toward Pittsburgh, and back along the B & O toward Somerset, church bells were ringing and hundreds of people were coming in from the country with their donations; and all day, one after another, relief trains kept streaming through, many of them with "For the Johnstown Sufferers" scrawled in big letters on the boxcars. One train in by the Somerset line carried a whole shipment of tents sent by the governor of Ohio. Another Pittsburgh train, eleven cars long, carried nothing but coffins.

Some of the offerings that were mounting up in Johnstown created more than a little amusement. In their eagerness to help, some people had not bothered to think much about what would be needed. One nicely tied bundle opened Sunday afternoon contained a ball of carpet rags, a paper of tacks, two bags of salt, one baby's shoe, and two darned stockings of different colors. A box of homemade liniment, with "warm before using" written on the side, was tossed out of one car. There was a package of worn-out schoolbooks, a Bible with several pages marked, some fancy needlework, even bits of bric-a-brac.

But almost everything else that came in, however shabby or

trivial seeming, was immediately grabbed up and put to good use. A blue dress coat with bright brass buttons that looked every bit of seventy years old was presented to an equally ancient-looking Grubbtown man who wore it away with much pride. Children went shuffling off in shoes several sizes too big for them. Women gladly put on men's coats and hats.

And as much as there was coming in, it was nowhere near enough. There were perhaps 27,000 people in the valley who had to be taken care of, who had to be supplied with every kind of basic necessity; and added to them were all those others streaming in to help.

By nightfall Sunday well over 1,000 people were in from out of town. Something like fifty undertakers had arrived from Pittsburgh. The railroad was bringing work crews in by the trainful. A Pittsburgh fire department had arrived and, remarkably enough, by midnight had just about extinguished the fire at the stone bridge. There was also present a rather stout Republican politician by the name of Daniel Hartman Hastings, the Adjutant General of the state, who, after looking the situation over since morning, had decided it was time the military took over.

A lawyer by profession, the general's only military experience had been at Altoona during the strike of '77. Saturday morning he had hitched up his team and driven nonstop from his home in Bellefonte, seventy miles to the northeast, arriving at Prospect Hill after dark. He had slept that night in the company of several tramps on the floor of the signal tower at the Pennsylvania station and managed to cross over to Johnstown first thing the next morning. He talked to Moxham and his committeemen about calling out the National Guard but was advised strenuously against it. Moxham thought it was important that the people handle their problems themselves; it would do more than anything else, he said, to help them get over their anxieties.

Later in the day, when a company of troops arrived from Pittsburgh, sent by the Pittsburgh Chamber of Commerce, Hastings told them to go back to where they had come from. They had received no proper orders to turn out, he said, and had no business being there. He gave the officer in charge a vigorous dressing down, and back they went.

But by nightfall another meeting was held with the local officials and it was agreed to draw up a formal request to the governor for troops. For by now it was clear to just about everyone that the job of running things had gone beyond what the Moxham "dictatorship" could cope with. In another two days Moxham would resign his authority altogether, and James B. Scott, head of the Pittsburgh Relief Committee, would take over as the civilian head.

Rumors of looting and drunken fist fights were now even more exaggerated than they had been the previous night, but now they were not totally unfounded. The Reverend Beale and others later testified to witnessing attempted thefts. On Prospect Hill there seemed to be an inexhaustible supply of whiskey. One husky farm boy who had come down from Ebensburg with a load of provisions stayed long enough to get so drunk that he toppled off the hillside, rolled head over heels down the embankment, and fell into the Little Conemaugh, nearly drowning in minutes. "God only saved him," his father said later, "and for something better we hope."

Captain Hart's police seemed unable to keep order, and if things were not troublesome enough as they were, one of his lieutenants, a much-respected local lawyer and sportsman named Chal Dick, went riding about on horseback brandishing a Winchester rifle and telling lurid stories about the Hungarians he had seen robbing the dead and how he had already shot a couple of them. The stories spread like wildfire, and with them went more fear and suspicion of any man who spoke with an accent or even looked slightly foreign. People talked of how Paris had been looted by the Germans during the Franco-Prussian War, or harked back to tales of violence and evil doings in the old country at the time of the plagues.

And to make matters still worse, it was well known that even more people were on their way. Word was Sunday night that Booth and Flinn, the big Pittsburgh construction company, was sending 1,000 men the next day, and everyone had heard about the kind of riffraff Billy Flinn was known for hiring. He would pack them into freight cars like cattle and then turn them loose into Johnstown. Every last one of them would have to be fed and sheltered, and where was it all to come from? And who was there to police the place?

But what was not known, even as Hastings sent out his message to the governor, was just how many others were heading for the devastated city. For along with the Flinn crew there were thousands more coming—charity workers, doctors, preachers, men looking for work, smalltime crooks and pickpockets, drifters, farm hands, ladies of the W.C.T.U., former Johnstown people heading back to look for relatives, railroad officials, prostitutes, sight-seers.

From Pittsburgh Captain Bill Jones was on his way with three carloads of supplies and a small army of 300 men from the Edgar Thomson works, a number of whom had been with him in the old days at the Cambria mills. In Philadelphia pretty society girls were packing medical supplies and making ready to start off with relief units organized by half a dozen churches. Mr. H. C. Tarr of the Utopia Embalming Fluid Company of Brooklyn had already struck out for Johnstown and would wind up traveling nearly 200 miles by horseback before he got there. In Washington, Miss Clara Barton and her newly organized American Red Cross had boarded a special B & O train.

For, already, the Johnstown Flood had become the biggest news story since the murder of Abraham Lincoln. On Saturday night, quite late, the reporters camped inside the brickworks had finally gotten a line clear to Pittsburgh, and the words had been pouring out ever since. ("The awful catastrophe at Johnstown is by all odds the most stupendous fatality ever known in the history of this country. . . .")

The news had an effect that is difficult to imagine; by Sunday it was spread across the front page of virtually every paper in the country. On Saturday the papers had hedged on how many had been killed. The New York *World* had reported 1,500 lives lost; the *Times* had been even more cautious, saying only that hundreds were dead. But on Sunday the *World* headlines ran halfway down the page, and though they still had no firsthand facts to go on, the editors had decided to pull out all stops:

10,000 DEAD

Johnstown Blotted Out by
the Flood

HALF ITS PEOPLE KILLED

Two Thousand Burned to Death in
the Wreck

ALL APPROACHES CUT OFF

In Pittsburgh the *Post-Gazette* was selling its editions so fast
that it had to reduce its page size temporarily in order not to run
out of paper. Everywhere people were talking of little else and
wanted to know more, much more; they wanted facts, names, details, pictures. And so along with all the others heading for Johnstown there came more reporters (perhaps a hundred or more), telegraph operators, editors, authors, artists, photographers.

The great rush to Johnstown, which had begun in Pittsburgh
Friday night, was now under way full force. They came, thousands
of them, from every station in life and from as far away as California, heading for a place very few of them had ever heard of two
days earlier, driven by the most disparate motives imaginable.

VIII

"No pen can describe . . ."

─────────

-1-

Henry S. Brown of the Philadelphia *Press* had been sitting at his desk at eleven o'clock Friday night when the news first came in. At 11:25 he was on board a westbound train pulling out of the Broad Street Station, having taken no time to pack or, for that matter, to give much thought to just where it was he was going or what chance he had of getting there. At Harrisburg the train was delayed by floods along the Susquehanna, but Brown stayed on board when the conductor assured him everything would be cleared up in a few hours and that they would be moving on again. At dawn he was told things had changed rather drastically; nothing would be open west for two weeks.

Brown got hold of some maps and decided that if he could get a train to Chambersburg, fifty miles to the southwest, and could hire a team there, he might just be able to drive the rest of the way, which, according to the map, looked to be another hundred miles. He took the Cumberland Valley Railroad out of Harrisburg, but it was not until Sunday afternoon that he reached Chambersburg, located a double team, and started over the Tuscarora Mountain to

205

McConnellsburg, twenty-two miles due west. Halfway over the mountain his wagon broke down, but he managed to borrow another from a farmer. At McConnellsburg he picked up another team and pushed on, along the Pennsylvania Pike (the old Forbes Road), heading for Juniata Crossing. From then on he splattered his way down washed-out roads, forded streams where bridges had been swept away, walked when he had to, crossed Sideling Hill in the dark, changed teams five more times, and never stopped to eat or rest. He reached Bedford about seven Monday morning and, finding no train there as he had expected, went whirling off once again, this time bound for Stoystown behind a pair of snow-white mules. Between Bedford and Stoystown, still traveling the state road, he managed to cross the Allegheny Mountain at a place where the elevation approaches 3,000 feet.

At Stoystown he would be able to pick up the B & O line from Somerset, but he arrived just in time to miss a relief train there; so rather than wait for another, he hired still one more team and headed on again, following the tortuous route of the raging Stony Creek down to Johnstown.

It was about seven thirty Monday night when he finally reached Johnstown, after having traveled the hundred miles from Chambersburg in about twenty-eight hours. No more than ten minutes later he was shaking hands with another correspondent by the name of F. Jennings Crute, also of the *Press*. Crute had left the Philadelphia office at the same time as Brown and had pulled into Johnstown only an hour before, having traveled about seven times as far as Brown. For instead of heading west on the Pennsylvania Friday night, Crute had made the whole trip by rail, first by heading east to New York, then going by way of Buffalo (on the Central) to Cleveland and Pittsburgh.

Brown and Crute went directly to work and, like the other reporters swarming over the place, were soon filing their stories from the brickworks above the stone bridge, which by now had become quite a center of operations. The Pittsburgh papers, the *Times*, the *Press*, the *Dispatch*, the *Commercial-Gazette*, and the *Leader*, were all represented. (In an old photograph taken at the end of the stone bridge on Sunday, a group of twenty-one Pittsburgh correspondents pose proudly beneath derby hats, several with cigar in hand, their dark vests crossed by heavy watch chains.)

They had taken over two floors of one building, as well as a wood-shed. The newcomers squeezed in where best they could, everyone working under tremendous difficulties. Those who had been there for more than twenty-four hours were unshaven, red-eyed, and near collapse from lack of sleep. They were using barrelheads, coffin lids, and shovel bottoms for writing desks, and the words they wrote were put on the wire as fast as was humanly possible.

The place became known as the "Lime Kiln Club" and rapidly gave rise to that special kind of fellowship-through-duress, which so often happens in war. "The culinary department," one of the group wrote later, "was taken charge of by Tom Keenan of the *Press.* With an old coffee-pot taken from the debris at the bridge, some canned corned beef, a few boxes of crackers, a few quarts of condensed milk and a bag of unground coffee, he was soon enabled to get up a meal for his starving comrades which was the envy of those in the neighborhood who, while hungry, did not belong to the band of scribes, whom they looked upon as a lot of luxurious revellers."

By late Monday the force of telegraph operators had increased enough to set up night and day shifts. Food became more plentiful, and the presence of the new men did much to boost spirits. The New York *Sun* reporters had come by the same roundabout route as F. Jennings Crute, while their rivals from the *Herald, World, Times,* and *Tribune* had gone more or less the way of Henry Brown. The correspondent of the Chicago *Inter-Ocean* walked up from Sang Hollow, as did several others, and every one of them was brimful of tales of his experiences.

The early arrivals at last got some sleep that night, there at the brickworks, while the newcomers found what accommodations they could elsewhere around town. Eight of them, including the Philadelphia men, wound up on the narrow first floor of the signal tower across the river. About midnight they were awakened by a man at the door saying, "Isn't this terrible. Look at them, human beings, drowned like rats in their hole." At which point one of the corpses sat bolt upright and said, "Get the hell out of here and let us sleep!"

But for all the boon companionship and oft-told stories, the hardships endured by "the gentlemen of the press" were considerable. Vile-smelling smoke from the still smoldering bridge blew through the windows of the old building where they worked. The floor was shaky and full of holes, and to enter the place in the dark of night was, as one man said, "to place one's life in jeopardy." John Ritenour of the Pittsburgh *Post* fell twenty feet, wedging between timbers and so severely injuring himself that he had to be sent home. Sam Kerr of the *Leader* fell off the top of a house lodged in the drift and would have drowned if one of his colleagues had not been on hand to pull him out. Clarence Bixby of the *Post* fell from the railroad bridge while trying to get across at one in the morning and was badly banged up. And several weeks later, F. Jennings Crute, worn down by lack of sleep and exposure, caught a cold that turned to pneumonia. On December 3 he died.

The competition between papers was friendly but fierce, with every man scrambling for an advantage. One of them, a William Henry Smith of the Associated Press, had actually been on board a section of the ill-fated *Day Express* and wrote a long, florid description of the experience. ("It was a race for life. There was seen the black head of the flood, now the monster Destruction, whose crest was raised high in the air, and with this in view even the weak found wings for their feet.") But for the rest it was a matter of finding out what was happening amid the chaos around them, and as of Monday night there was plenty happening.

The city itself was still the most overpowering spectacle. ("It is a scene that blanches the faces of strong men, and in its multiplying horror is almost beyond description," wrote a reporter for the New York *Daily Graphic*.) The weather had turned dull and cold again, which was unpleasant but welcome news as far as the doctors and sanitation workers were concerned. This way the dead would not decay quite so fast.

Bonfires by the hundreds were blazing across the valley where the ungainly and by now putrid carcasses of drowned horses were being cremated. The stench everywhere was terrible, of burned plaster and sodden bedding, of oil-soaked muck, of water thick with every kind of filth, and, worst of all, of still unfound bodies. The correspondents wrote of negotiating the rope bridge over the Conemaugh ("A slide, a series of frightful tosses from side to side, a run, and you have crossed . . .") and of the curious things to be found once in town ("In the midst of the wreck a clothing store dummy, with a hand in the position of beckoning to a person, stands

erect and uninjured."). They interviewed bystanders (" 'I have visited Johnstown a dozen times a year for a long time,' said a businessman to-day, 'and I know it thoroughly, but I haven't the least
idea now of what part of it this is. I can't even tell the direction the
streets used to run.' "); and they quoted General Hastings as saying
that there were 8,000 people dead. ("Nobody thinks this too
small," the *Sun* reporter added. "Nobody who has been about here
an hour would think anything too awful to be possible.")

Sunday night four enormous relief trains had rolled in below
the bridge. Monday Billy Flinn brought in 280 teams of horses and
1,300 men. ("Very few Americans among them," wrote one reporter.) Mrs. Lew Wallace, wife of the war hero and novelist, was
reported missing from the *Day Express*. (She had actually taken
another train and was safe in Altoona.) John Fulton and Colonel
John Linton were both mistakenly reported dead (Fulton was
reported "positively drowned"), and James McMillan, vice-
president of Cambria Iron, was asked when work would start on
rebuilding the mills, to which he answered, "Immediately." There
was talk of dynamiting the wreckage at the stone bridge, and there
was a strong plea from the doctors and the sanitation officials from
Pittsburgh to let it burn. The smell of burning flesh among the
wreckage was something awful ("People in New York who remember the smell of the ruins of the Belt Line stables, after their
destruction by fire . . . know what the odor is."), but fire would
cut the odds against a typhus outbreak, and throughout the valley
and on downriver, clear to Pittsburgh, typhus had become an overriding concern.

In Pittsburgh the papers urged everyone to boil his water.
From Nineveh, where nearly a hundred bodies had been recovered,
Dr. Benjamin Lee, head of the Pennsylvania Board of Health, sent a
message to the sheriffs of the four counties between Johnstown and
Pittsburgh:

The State Board of Health hereby directs and empowers you
to immediately summon a posse to patrol the Conemaugh river,
tear down the drift heaps and remove the dead bodies, both
human beings and domestic animals. This is absolutely necessary to protect your county from pestilence.

The wreckage at the bridge was described in detail, with some saying it covered thirty acres, others claiming it was more like sixty. (It was about halfway in between.) "I stood on the stone bridge at 6 o'clock," wrote a *Sun* reporter Monday, "and looked into the seething mass of ruin below me. At one place the blackened body of a babe was seen; in another 14 skulls could be counted . . . At this time the smoke was still rising to the height of 50 feet . . ." On Wednesday, June 5, a little boy named Eddie Schoefler would be found still alive amid the wreckage. It would be one of the momentous events of the week.

Then, from Sunday on, there had been increased tension over the Hungarians, which was something quite colorful indeed to write about. Thanks to Chal Dick and, by now, many others, tales of "foul deeds" perpetrated by the "fiendish Huns" were rampant, and only a few reporters bothered to try to check them out. Story after story went on the wire describing how "ghouls, more like wild beasts" were slicing off fingers for gold wedding bands, and how angry Johnstown vigilantes were hunting them down. One account described how a woman's body had been decapitated in order to steal her necklace. The *Post* told how gangs of Hungarians tried to raid unguarded freight cars for food and clothes. Another report said that a Hungarian had been caught in the act of blowing up a safe in the First National Bank. The *Daily Graphic* described how a crowd cornered a Hungarian at his "fiendish work" and strung him up on a lamppost.

This sample of the over-all tone and content of the reports was written late Sunday:

> Last night a party of thirteen Hungarians were noticed stealthily picking their way along the banks of the Conemaugh toward Sang Hollow. Suspicious of their purpose, several farmers armed themselves and started in pursuit. Soon their most horrible fears were realized. The Hungarians were out for plunder. They came upon the dead and mangled body of a woman, lying upon the shore, upon whose person there were a number of trinkets of jewelry and two diamond rings. In their eagerness to secure the plunder, the Hungarians got into a squabble, during which one of the number severed the finger upon which were the rings, and started on a run with his

fearful prize. The revolting nature of the deed so wrought upon the pursuing farmers, who by this time were close at hand, that they gave immediate chase. Some of the Hungarians showed fight, but, being outnumbered, were compelled to flee for their lives. Nine of the brutes escaped, but four were literally driven into the surging river and to their death. The thief who took the rings was among the number of the involuntary suicides.

The "thugs and thieves in unclean hordes," as one writer described them, were nearly always Hungarians, though there was at least one report of two Negroes being shot at by Pittsburgh police when seen robbing a dead body, and there were a few references to "the worthless Poles."

Such accounts were given a great deal of space by all but a few of the big eastern papers and were featured prominently in the headlines. ("FIENDS IN HUMAN FORM" ran the New York *Herald* headline on Monday. "DRUNKEN HUNGARIANS, DANCING, SINGING, CURSING AND FIGHTING AMID THE RUINS.") Lurid illustrations were published, drawn by artists who had only the reporters' stories to go by. One scene showed two bodies dangling from a telephone pole near the riverbank, while in the foreground a "wild-eyed" Hungarian, who looks much like a touring company Fagin, is held at bay, knee-deep in water, by a stalwart gentleman with a horse pistol who could very well be Robert E. Lee.

They were stories which had great appeal to anyone ready to believe in the darker side of humanity and particularly that segment of humanity which spoke with a thick accent, smelled of garlic, and worked cheap. The only trouble was that there was scarcely any truth to the stories, as several correspondents had already begun to suspect. At four Monday afternoon Alfred Reed of the *World* cabled his editors:

NO LYNCHINGS. I WARNED YOU LAST NIGHT NOT TO PRINT WILD RUMORS, AND AM GLAD TO HEAR YOU HAD ENOUGH CONFIDENCE IN ME TO HOLD OUT SUCH STORIES.

The next day an angry General Hastings issued a statement that reports of lynchings and rioting were "utterly devoid of truth," sharply criticized the newspapers for publishing them, and suggested that the reporters stick to the facts.

Several characters had indeed been caught trying to pilfer the dead and had received some rather rough treatment, including, it appears, enough mock preparations for a lynching to put a terrific scare into one of them; and it is quite possible that a few fingers may have been mutilated by thieves trying to wrench off gold wedding bands. But there were certainly no diamond rings stolen (one survivor doubted that there were more than one or two diamond rings in all Johnstown at that time), no bank safes were blown, and, as David Beale wrote later, no fingers were cut off by human ghouls. Furthermore, the Hungarians themselves apparently had almost nothing to do with what foul doings there were. "There was little stealing done by the Hungarians," Beale wrote, "and most accounts of outrages attributed to these people were apocryphal; and I am glad to say that all statements of shooting and hanging them were without foundation." And to emphasize the validity of this last statement, he said his source was Chal Dick himself.

Dick, it seems, had gone slightly out of his head immediately after the disaster and had been suffering from vivid and vicious delusions. His wife and children had been killed and he simply went berserk for about a day or so. By the time he snapped out of it, the damage had been done, and from then on the stories were spread, according to the best evidence, largely by outsiders who had come into the valley.

For though there may have been relatively little resentment in Johnstown against the Hungarians (or the other Southern European peoples called Hungarians), in Pittsburgh feelings were different. The steel bosses, like Henry Clay Frick, had been bringing them in by the thousands to work in Braddock and Homestead. They were single men mostly, willing to work for the lowest wages, and under the worst conditions, just to save enough to go back home and buy a small farm on the Danube. They got the toughest jobs, worked hard, and were generally hated by the Irish, the German, and American workers. Years later, John Fitch, the historian, interviewed an old Scotch-Irish furnace boss in Pittsburgh about the "hunkies."

"They don't seem like men to me hardly," he said. "They can't talk United States. You tell them something and they just look and say, 'Me no fustay, me no fustay,' that's all you can get out of 'em." When wages were going down, when men were let go at the mills, when the unions suffered setbacks, somehow the Hungarians seemed at the root of things.

The few Hungarians there were in Johnstown (perhaps 500 of them were living in the valley at the time of the flood) were subjected to days of abuse. Speaking little English, fearful and suspicious even under normal circumstances, they now became so terrified of the angry crowds that hung outside their homes that they dared not go out even to collect their share of the relief provisions. Their children were starving; the men grew desperate. At one point about twenty of them were encouraged to come out to help dig graves in the cemetery above Minersville. After working all day, on their way back home in the dark, they were set upon by a gang armed with clubs and were badly beaten.

But by midweek the Hungarian scare was over. There were still rumors, but papers like the Philadelphia *Press* were saying, "There is not an inch of truth in them," and nearly everyone in Johnstown knew that that was so. While in Chicago the *Herald* wrote that the "Magyars" there were "justly indignant" over the stories, and tried to resolve the whole unfortunate business by adding, "The wretches who now prey upon the dead at Johnstown and refuse to aid in the work of rescue, are undoubtedly Bulgarians, Wallachs, Moldavians, and Tartars, classes degraded in all their manners as is the North American Indian. . . ."

Sometime Monday Colonel Unger came into town from South Fork, accompanied by the Shea brothers, John Parke, and one or two other employees of the fishing and hunting club. Understandably, the press was most interested in talking to them.

Unger gave the Pittsburgh *Post* a brief rundown on what had happened at the dam Friday morning and how he and his men had tried to prevent the disaster. He estimated that the loss to the club was about $150,000 and said that the club members who had been at the lake were all safe and that they had gone off to Altoona.

Parke seemed more interested in getting word to his people in Philadelphia that he was alive and in good health, but was quoted by the New York *Sun*, "No blame can be attached to anyone for this greatest of horrors. It was a calamity that could not be avoided." He said the fault was "storm after storm" and that "by twelve o'clock everybody in the Conemaugh region did know or should have known of their danger."

But an employee by the name of Herbert Webber, who must have been interviewed separately, launched into a long description of the dam the morning before it failed. He told the reporters that at around eleven he had been attending to a camp a mile back from the dam when he noticed that the surface of the lake seemed to be lowering. He could not quite believe what he saw, he said, so he went down and made a mark on the shore, and sure enough he found his suspicions were well founded. For days before, he went on, he had seen water shooting out between the rocks on the front of the dam, so that the face "resembled a large watering pot." The force of the water was so great "that one of these jets squirted full thirty feet horizontally from the stone wall." When he ran up to the dam that morning, he declared, he saw the water of the lake "welling out from beneath the foundation stones."

The story was preposterous, of course, and had no connection with what actually happened, but the reporters had no way of knowing that. The watering-can image made splendid copy, so out it went, along with everything else.

But by this time at least one enterprising reporter had already made his way to the club. In Pittsburgh that Monday the headline across the front of the *Post* read: "TO THE DAM AT LAST." The story had been sent out at nine the night before and said, as Unger had, that the Pittsburgh people were safe and, as Parke had, that warnings had been sent down the valley before the break. In another two days more reporters would show up at South Fork. They would begin looking over the construction of the dam itself and start questioning the local people about the club. South Fork would shortly become the center of a stormy series of events, but for now Johnstown remained the major focus of attention.

One survivor after another was interviewed and dozens of frightful personal experiences were penciled into reporters' notebooks. The heroism of Bill Heppenstall (Hepenthal several papers spelled it), the adventures of Gertrude Quinn, and John Hess's ride into East Conemaugh were described at length.

Numerous stories were collected of ironic or incredible things that happened. All of Johnstown's three or four blind people had survived the flood. Frank Benford's dun-colored mare was found in an alley next to where the Hulbert House had been, up to her belly in debris, alive, but blinded in both eyes. Old Mrs. Levergood, widow of Jacob Levergood, whose father had owned the town way back in the early days, was found dead, all the way up at Sandy Vale, still seated in her rocking chair.

Then there was the story about the engineer at the Cambria works who early in the afternoon of that fateful Friday had started a letter to an old college friend, "Thank God, I am through with a day such as I hope never to pass again." A "gay girl" of the town was said to have jumped from a hotel window during the very worst of the flood in a fatal effort to save a drowning child. There was a Newfoundland dog that supposedly hauled a Woodvale woman to safety and then swam back to save a drowning baby. And another dog, a water spaniel named Romeo, was said to have towed his mistress, Mrs. Charles Kress, to the windows of Alma Hall. And just above the hideous pileup at the stone bridge, on a billboard at the depot, there was a large poster, undamaged by the flood, which several reporters made a point of mentioning. Put there a few days before the flood to announce the arrival of Augustin Daly's *A Night Off*, its very large headline read, "Intensely Funny."

Among the best pieces describing the human condition in Johnstown during these days were several by a late-arriving cub reporter for the Philadelphia *Press* whose name was Richard Harding Davis. The son of two prominent Philadelphia literary figures, he was strikingly handsome, twenty-five, elegant, aloof, and loving every minute of his first real assignment. At Lehigh he had been the most popular figure on campus, despite what one of his classmates described as his "strict adherence to everything English in the way of dress and manner." His first job had been on the Philadelphia *Record* where he sported a long, yellow ulster, carried a cane, and was rather hard for the old newspapermen to swallow. As it was, he only lasted three months. He was caught one day by the city editor writing up an assignment with his kid gloves on and was promptly fired on the spot.

But with the *Press* he had fared better. He had sold a few

short stories, interviewed Walt Whitman, and sent some samples of his work off to Robert Louis Stevenson with a request for advice. Stevenson advised writing "with considerate slowness and on the most ambitious models." The slowness Davis would never quite master, nor would he try really; and the only ambitious model he ever seems to have set his sights on was his own very clear picture of himself as the world's most dashing and celebrated foreign correspondent.

At the time the Johnstown story broke, Davis had been on vacation. It took several days for him to persuade the paper to send him, and when he finally arrived, he got off to a characteristic start. "A Philadelphia reporter was sent here to finish up the disaster, but the disaster is likely to finish him," wrote *The New York Times* man. On alighting from his train, Davis "paralysed Newspaper row" by asking for the nearest restaurant. When it was explained that everyone had to forage on the country, Davis wanted to know where he could hire a horse and buggy, which set off another round of laughter. "But," concluded the *Times* man, "he capped the climax by asking where he could buy a white shirt. A boiled shirt here is as rare as mince pie in Africa."

Boiled shirt or no, he went to work, concentrating on human-interest stories. He wrote of walking over thousands of spilled cigars and of a pretty, young relief worker named Miss Hinkley of Philadelphia who was

> . . . sitting busily writing at a table beside an open window which looked out on the yard of the morgue, and in which forty odd coffins filled with the dead were being examined by the living. Miss Hinkley's hair was as carefully arranged and her tailor-made gown as neat and fresh as if she had stepped that moment from the Quaker City's Rittenhouse Square. Reporters became painfully conscious of clothes that have been slept in for seven nights, and chins that had forgotten razors.

He wrote about a fist fight down in Cambria City between a local deputy sheriff and a drunken National Guard lieutenant named Jackson, who was put under arrest and sent back to Pittsburgh after several bystanders, including Davis, stepped in to break things up. He described the offers coming in from people all over the country who wished to adopt a Johnstown orphan, and told the story of a man named John McKee, whose body had been found inside a cell of the town jail. McKee had been locked up for twenty-four hours for overcelebrating on Memorial Day.

And along with the reporters, working their way among the ruins, came the photographers, lugging their ponderous, fragile equipment. They made pictures of men standing on freight cars in the midst of Main Street, of wagons loaded down with coffins, and of the great barren mud flat where Woodvale had been. The monstrous debris that clogged the city was pictured from virtually every angle, and at least one photographer decided to improve slightly on his composition by having a man lie down and play dead in the foreground. The picture later became one of the most popular stereoptican views of the disaster. But by the time the photographers were about, any body so exposed would have been long since found and removed; moreover, the shirt on the man's back looks a bit too neat and clean and the things around him are a little too nicely arranged.

Upturned houses, gangs of laborers carrying shovels and axes and threading their way through huge dunes of rubbish, like a drab, derby-hatted army moving through the remains of a fallen city, the jerry-built shelters on the hillsides, farm women in poke bonnets working at the commissaries, they all made splendid subjects. But the most popular subject by far was a house owned by John Schultz which had stood on Union Street but was now stranded at the east end of Main. It had been pitched up on its side, and through an upstairs window a gigantic tree had been driven, its roots jutting thirty feet into the air. The building looked as though it had been skewered by some terrible oak-flinging god. One by one men and boys would crawl out on the tree to sit for a portrait, their faces registering no emotion, their feet dangling in light that had little more color than it would have in the final printed photograph. Six people had been in the house when the water struck, and they had all come out alive.

At one point it was estimated that there were no less than 200 amateur photographers about town, enough in any case that they had become a serious nuisance. So the word went around that if you were an able-bodied man but had no official business in town, then you had to work if you wanted to stay on. It was, as one observer said, a policy which had a "most salutary effect."

Harper's Weekly, Frank Leslie's, and some of the other picture magazines had sent artists to cover the story. There were two or three writers gathering material for quick books. And in New York the *World* even managed to get Walt Whitman, who had celebrated his seventieth birthday on the day of the flood, to write a poem which was promptly printed on page one.

> A voice from Death, solemn and strange, in all his sweep
> and power,
> With sudden, indescribable blow—towns drown'd—humanity
> by thousands slain,
> The vaunted work of thrift, goods, dwellings, forge, street,
> iron bridge,
> Dash'd pell-mell by the blow—yet usher'd life continuing
> on, . . .

And in the Denver, Boston, and Brooklyn papers, long excerpts were published from a novel called *Put Yourself in His Place,* which had been written by a well-known English author named Charles Reade nearly twenty years before. Its closing chapters described the bursting of a reservoir and a dreadful flood which were surprisingly similar to what had happened at Johnstown.

Reade had based his book on the failure of the Dale Dyke at Sheffield, England, which had taken 238 lives in 1864; but for the millions of Americans who now read the excerpts, the "Hillsborough" of his story, with its steel mills and coal mines "fringed by fair woods," and its reservoir in the mountains to the east, seemed so like Johnstown as to be uncanny. The story told of people dreaming of floods, of workmen who had long "misliked" the foundation of the dam. When the break came, it was only after a storm had raised the level of the lake so far that it started flowing over the center of the dam. Then down the valley came "an avalanche of water, whirling great trees up by the roots, and sweeping huge rocks away, and driving them, like corks, for miles."

Meanwhile, the headlines blared away, day after day. "Agony" . . . "WOE!" . . . "PESTILENCE!" . . . and by midweek,

"DEATH GROWS—A GIANT! One Pervading Presence Throughout the Conemaugh Valley, FIFTEEN THOUSAND CORPSES, A Tale of Grief That Can Only Be Told in Bitter Tears, Another Day of Utter Despair."

The phrase "no pen can describe . . ." kept cropping up again and again, but the pens kept right on describing. The story took up the entire front page of both *The New York Times* and the *World* for five straight days. The Boston *Post* carried little else on its front page for twelve days running. It was called "The Great Calamity," "The Nation's Greatest Calamity," "The Historic Catastrophe." *Frank Leslie's* said outright, "It is the most extraordinary calamity of the age." Great battles had destroyed more life, said one writer after another, but no battle left such a ghastly trail of horror and devastation. That such a thing had happened in the United States of America in the year 1889 seemed almost more than the editorial writers could accept. Several papers, including *Frank Leslie's*, allowed that similar slaughter might occur in India or China or other remote lands "where human life is cheap," but how in the world had it ever happened here?

All over the country newspapers published column after column of names of the dead. An extraordinary amount of space was given over to telling of the "Slaughter of the Innocents." The bodies of women and children found among the wreckage or in the deep, flood-dumped silt downstream were described in grim detail. Every reporter at Johnstown it seems saw at least one dead mother still clutching her dead child, and much was made of the fact that more women died in the disaster than men. (Of the bodies finally recovered 923 were men, 1,219 women.)

Victorian sentimentality had a heyday. The most pathetic-looking "Johnstown orphans" imaginable were drawn by New York artists and published beside long accounts of lost children. There were stories published of families tenderly bidding each other a final good-by just as the flood was about to pounce on them and of people tucking farewell notes into bottles before they slipped beneath the water for the last time.

One publisher, Kurz & Allison of Chicago, eventually got out a color lithograph which became one of the popular works of art of the age. In the upper right-hand corner a dam bursts almost directly on top of a Johnstown where all the women and children are barefoot and many are in their night clothes at four in the afternoon. The women are fainting, falling, down on their knees praying, while the men, most of whom are amply clothed and shod, dash about trying bravely to contend with the rush of fire and water.

The newspapers, too, went very heavy on the horrors to be seen among the ruins of Johnstown, sometimes stretching the art nearly to the breaking point. This memorable sample, written at Johnstown on Wednesday, June 5, appeared the following day in the Philadelphia *Press:*

> . . . One of the most ghastly and nauseous sights to those unaccustomed to scenes of death is the lunching arrangement for the undertakers. These men are working so hard that they have no time for meals, and huge boilers of steaming coffee, loaves of bread, dried beef and preserves are carried into the charnel house and placed at the disposal of the workers. Along comes one weary toiler, his sleeves rolled up, and apron in front and perspiring profusely despite the cold, damp weather. He has just finished washing a clammy corpse, has daubed it with cold water, manipulated it about on the boards and in the interval before the body of another poor wretch is brought in, gets a cup of coffee and a sandwich. With dripping hands he eats his lunch with relish, setting his cup occasionally beside the hideous face of a decomposing corpse and totally oblivious to his horrible surroundings.

Whatever the reporters may have lacked in the way of facts, they made up for in imagination. Distortions, wild exaggerations, and outright nonsense were published in just about every major paper in the country. One reporter described buzzards ("incited by their disgusting instinct") circling over the stone bridge; another claimed the rivers were literally dammed with dead bodies. There were tales of wild dogs ravaging the graves of flood victims and devouring corpses by the dozens. Indianapolis readers were told that "each blackened beam hides a skull."

There seems little doubt that there was plenty of drinking going on, but one writer for the New York *World* had men staggering about with whole pailfuls of whiskey. ("Barrels of the stuff are constantly located among the drifts, and men are scrambling over each other and fighting like wild beasts in their mad search for it.") A large family was pictured sailing by during the height of the flood singing "Nearer, My God, to Thee" in harmony. And of those stories emphasizing the pitiful fate of the innocent, perhaps the most imaginative was one filed from New Florence. A bride from Johnstown (she was supposedly married just before the flood) was quoted: "Today they took five little children out of the water, who had been playing 'Ring around a rosy.' Their hands were clasped in a clasp which even death did not loosen, and their faces were still smiling."

But the most splendid story of the lot was one about a man named Daniel Peyton, the so-called "Paul Revere of the Flood," who was said to have galloped down the valley on a big bay warning everyone to run for the hills. Peyton (in some versions he is Periton) seems to have evolved out of young John Baker of South Fork and John Parke, Jr. He can also be traced to Charles Reade's *Put Yourself in His Place*, where a rider sped through the night warning isolated farm families that the water was on the way. In most versions of the Peyton story, including the best-known of several epic poems, *The Man Who Rode to Conemaugh*, by John Eliot Bowen, which was published first in *Harper's Weekly*, the hero gallops the length of the valley just ahead of the onrushing wave. (The fact that there was no valley road on which to make such a ride never seemed to bother any of the authors very much.) Pale, his eyes aflame, he cries out "Run for your lives to the hills!"—then dashes on.

> Spurring his horse, whose reeking side
> Was flecked with foam as red as flame.
> Wither he goes and whence he came
> Nobody knows. They see his horse
> Plunging on his frantic course,
> Veins distended and nostrils wide,
> Fired and frenzied at such a ride.

Nobody pays any attention to him. They decide he is a lunatic and jokingly dismiss the whole thing.

"He thinks he can scare us," said one with a laugh,
"But Conemaugh folks don't swallow no chaff;
'Taint nothing, I'll bet, but the same old leak
In the dam above the South Fork Creek."

In one version published as a Sunday-school lesson in 1891, the messenger goes into a saloon to spread the warning and winds up getting so drunk he can go no farther. But in either prose or poetry, in the end the flood finally catches up with poor Peyton, and nowhere is the moment more superbly described than in a book titled *The Official History of the Johnstown Flood*, which was put together that June by two Pittsburgh newspapermen named Frank Connelly and George C. Jenks. Peyton, in their "official" account, is about thirty years of age, and though he does not come clear from South Fork, but sticks only to the streets of Johnstown (Connelly and Jenks apparently felt they could not quite get away with the valley ride), his message and manner are essentially the same, and his end comes this way:

. . . At last he completed the circuit of the city and started in search of a place of safety for himself. To the hills he urged his noble steed. Tired out from its awful ride, the animal became slower and slower at every stride, while the water continued to come faster and faster in pursuit. Like an assassin upon the trail of his victim, it gained step by step upon the intrepid rider. But the hills are in sight. No, he is doomed, for at that moment a mighty wave, blacker and angrier than the rest, overtook horse and rider and drew both back into the outstretched arms of death.

As the Reverend David Beale wrote, "This fate was very necessary to the story, as it rendered an interview of the hero by another impossible."

Though the story appeared in a dozen or more different versions and was accepted outright as fact, it was quickly discredited in Johnstown. For there seemed to be no one who actually saw this Daniel Peyton. Furthermore, as near as anyone knew, and according to every available record, there never was anyone by that name in South Fork, Johnstown, or any other place in the valley. Victor

Heiser and some of his friends got so interested in the tale that they spent some time trying to track it down. If there were such a fellow, they wanted to meet him; and if he had been killed as people said he had, then they surely wanted to see him get the credit he had coming. But they never turned up anything, nor did anyone else.

Still, with all the stretching of facts, with all the fabrication and bunk being printed, no one seemed to mind very much. If the horror of what had happened was not described exactly according to facts, people knew that what had happened was still a great deal worse than any words could convey, however accurate. And if a few small fables had been called up for the occasion, well they were really no more extraordinary than a dozen other stories that were "the God's truth."

For the publishers it was one of the headiest weeks ever. Newspaper circulations broke all records. For days on end, one edition after another sold out almost as soon as it hit the streets. The New York *Daily Graphic* was selling an unheard of 75,000 copies a day. In Pittsburgh there seemed no letup to the clamor for more news. A new weekly picture newspaper called the *Utica Saturday Globe*, published in upstate New York but widely circulated, increased its circulation by better than 63,000 with its special addition on the disaster.

Songs were written, including one called "Her Last Message," which was more or less based on Hettie Ogle, and another called "That Valley of Tears," which was about a baby who was swept from the hearthside and drowned in its cradle. The last one, arranged for piano and orchestra, closed with the lines:

> And there midst all that wreck, with cruel waters laved,
> That babe within its cradle bed tho' dead 'tis saved;
> Saved from a life of toil and worldly care,
> Oh! That we could in thy glorious prospects share.

Magazines such as *Harper's Weekly* and *Frank Leslie's Illustrated Newspaper* got out special editions filled with pages of pictures and maps. Books were dashed off in a few weeks and rushed to the printers. Before the year was out, in less than six months, a dozen books would be written and published, most of them little more than an assembly of newspaper accounts, full of repetition, contradictions, and abundant nonsense. Several became best-sellers. One, a rousing period piece called *The Johnstown Horror*, was on sale in Johnstown itself in less than a month after the tragedy.

-2-

For Johnstown the result of the journalists' handling of the story was even more staggering. The enormous sympathy aroused by the newspaper accounts, the pictures, the songs and poems, brought on the greatest outpouring of popular charity the country had ever seen. (And this too, alas, the journalists felt obliged to immortalize: "As the bow of promise gilded the Oriental sky after the Noachian deluge, so the dark cloud enfolding the Conemaugh Valley had a ray of brightest sun light. A great, grand glorious tide of sympathy for the sufferers swept the land like a conflagration, warming men's hearts to deeds of radiant luster.")

On the Saturday following the calamity more than $100,000 had been raised in Pittsburgh. By the time they had finished, the people of Pittsburgh would give $560,000. New York City gave $516,000; Philadelphia, $600,000; Boston, $150,000. Nickels and dimes came in from school children and convicts. Churches sent $25, $50, $100. In Salt Lake City thousands of people turned out for a concert given in the huge Mormon Tabernacle, the proceeds of which were sent to Johnstown.

In New York's Madison Square Garden, Jake Kilrain, who was to take on John L. Sullivan in another few weeks for the world championship, put on an exhibition fight with Charley Mitchell to raise money for Johnstown. (Sullivan was invited to attend the show but did not, and was soundly hissed when his telegram of apology was read by the manager.) At the Metropolitan Opera House, Edwin Booth played the third act of *Othello* and raised $2,500. In Washington, John Philip Sousa gave a band concert. In Paris, Buffalo Bill staged a special production of his Wild West Show, which was attended by the Prince of Wales. And in Altoona, on the night of Monday, June 3, there had been a benefit performance presented by the *Night Off* troupe.

Tiffany & Company sent $500. R. H. Macy & Company sent $1,000. Joseph Pulitzer sent $2,000; Jay Gould, $1,000; John Jacob Astor, $2,500. The New York Stock Exchange gave $20,000. An old Confederate soldier sent four $100 Confederate bills, and the citizens of Cupola, Colorado, sent a solid-silver brick.

There were donations from Nantucket ($1,136.93), Yazoo City, Mississippi ($350), and Tombstone, Arizona ($101). The first check to arrive was supposedly one for $100 from Senator Matthew S. Quay, leader of the Republican forces in Pennsylvania; and Simon Cameron, the state's crusty, old Republican boss, who would be remembered as the man who defined an honest politician as one "who when he is bought stays bought," sat down and wrote out what was said to be one of the last checks he ever signed, for $1,000.

The Hebrew Benevolent Society of Los Angeles contributed $1,000. The United States Brewers Association sent $10,000. The Pittsburgh Society of Spiritualists collected $100. Money poured in from every state and from fourteen countries overseas. The London Stock Exchange gave $5,000. The total donations from Germany came to $30,000. There was money from the Lord Mayor of Dublin, the Mayor of Belfast, and the Sultan of Turkey. Queen Victoria sent her condolences to President Harrison, and from Washington came more than $30,000, including a check from Harrison for $300. (The President had presided over a mass meeting at the Willard Hotel Tuesday afternoon, looking very small and gray as he sat in a big armchair in the center of the stage. He had made a brief appeal for help, during which, according to one account, his voice trembled, and nearly $10,000 had been raised.)

In all, the contributions from within the United States would come to $3,601,517.80. The sum from abroad was $141,300.98, making a total of $3,742,818.78. And this does not include the goods of every kind that rolled in by the trainload.

From Saturday on the relief trains kept coming without letup. Lumber came in by the carload, and furniture, and barrels of quicklime, embalming fluid, pitch, pine tar, and resin. There was a carload of potatoes from Walla Walla. Minneapolis sent sixteen carloads of flour. Cincinnati sent 20,000 pounds of ham. Prisoners at the Western Penitentiary baked 1,000 loaves of bread. From

Arbuckle's in Pittsburgh came 30,000 pounds of coffee, and a New York butcher sent 150 pounds of bologna. Wheeling, West Virginia, sent a whole carload of nails.

By Friday, June 7, two hundred carloads of provisions had cleared Pittsburgh. At the Pennsylvania depot in Johnstown, and at the B & O depot, the platforms and yards were piled with cans of biscuits, boxes of candles, cheese, lamp chimneys and matches, huge cases of soap and canned goods, bacon by the barrelful, and hundreds of sacks of corn meal. People had donated cots, mattresses, hair combs, pipes, pillows, teakettles, tents, cookstoves, and more than 7,000 pairs of shoes.

By Thursday, the 6th, the day the Cincinnati hams arrived, any fears there had been of serious food shortages were over. The problem now was distribution. Commissaries had been set up and were run with reasonable efficiency. Several women complained that the prettiest girls were getting more than their share, and often people came climbing back up the hillsides with empty baskets after waiting hours in line. But considering how many more there were to feed, things were exceptionally well organized. The Army had arrived the day before, which helped a lot. Hastings had held off calling in more troops but had decided late Tuesday to bring in the 14th Regiment from Pittsburgh (580 men), and now with their white tents pitched in rows where the park had been and on Prospect, the place began to look like a cross between a military encampment and a wide-open mining town. For along with the Army had come another Booth and Flinn gang of 1,100, and with the railroads bringing in their own crews, with more and more drifters arriving, by the end of the week there would be nearly 7,000 men at work in the valley. They too pitched tents or built rough shanties on the hillsides, upon which they hung names like "Lively Boys," "Willing Workers," and "The Ladies' Pets." They were making $2.00 a day, which was good money; and a man with a team could make $5.00, which was good enough to have attracted men all the way from Ohio and New York. And with the pay went "room and board."

Enormous tents served as so-called dining halls, where several hundred men could be fed at a sitting. Coffee was ladled out of a bucket, bread was passed out in big dishpans, followed by more dishpans, each with a ten-pound slab of butter in it. When meals were ready, "the rush for the table cast the Oklahoma boom all in the shade," wrote the *Times*.

The men worked hard and made notable progress. By sundown Thursday more had been accomplished than in all the days since the disaster. The air rang with shouts and the screech of nails yanked out by crowbars. Hundreds of bonfires crackled and sent up columns of black smoke where now barrels of resin were being dumped on to burn up dead animals and the worst of the debris. Teams were hitched to mud-bound machinery and dangerous-looking walls that came down with a crash. There were tools enough now for every man, and by midweek a shipment of dynamite along with an expert to handle it had arrived to break the jam at the stone bridge.

So far every attempt to pull loose the debris there had failed. Locomotives and a steam winch had been tried, and a gang of lumberjacks from Michigan had done their best but had made hardly a dent. The ugly mass had been driven against the bridge with such savage force, and was so tightly ensnared by miles of barbed wire, that nothing, it seemed, could break it loose. Then for several days the valley echoed with the roar of dynamite as slowly, one by one, great gaps were blasted through the entanglement, which, most people believed, still held many human corpses.

The expert in charge was a man named Kirk, who was known among the press as "The Prince of Dynamiters." He was short, squat, about fifty years old, with a grizzly beard and a habit, when walking, of waving his hands about as if warning people back from his next explosion.

"He personally superintends the preparations of all blasts," wrote the *Times* man on the scene, "and when ready emits a peculiar cry more like a wail than a warning. Then he surveys the atmosphere with the air of a Major General and yells, 'Fire!' The yell often terrifies the spectators more than the explosion itself." But on Saturday Kirk managed to scare the daylights out of everybody, and with a good deal more than his yell.

Despite the inroads he was making, he was not at all satisfied with the time the job was taking, so he cut loose with a 450-pound charge, hoping to break the jam with one blow. Nine fifty-pound boxes of dynamite were planted at the base of the mass, each set about thirty feet apart. The whole valley and every building that was still standing trembled with the blast. Horses shied violently. Work stopped all over town. And within no time Kirk had been told in quite explicit language by what seemed like half the surviving population that he was to pull no such stunt again. But a gaping hole had been gouged out of the wreckage and at last daylight could be seen under the arches.

Some men became so exhausted from the work that they had to give up and go home. Everyone by this time had grown more or less accustomed to the grisly look of the place, the smell, and the constant presence of death; but the work itself was grueling, and no matter how much was accomplished, there seemed always an insurmountable lot still left to be done.

The wreckage spewed across the city was far greater in quantity than all that had stood there before the flood. The water had swept the valley above the city of virtually everything that man had built there over some eighty years and dumped it on Johnstown. Now every last bit of it had to be cleaned up, searched through, burned, or carted away. On top of that, sanitation problems were enormous. One gang of workers did nothing but sprinkle disinfectants over the entire district, and as George Swank later wrote, there was "a fortune for the man who concocts a disinfectant that won't stink the nose off a person." Four thousand barrels of quicklime would be used before the cleanup was finished.

Another gang was detailed to gather and burn bedding, clothes, and carpets which had been ruined by the water. The houses that had not been swept away were left, as Dr. Matthews said, in the most unsanitary condition imaginable. "The flood water," he wrote, "was heavily charged with every kind of filth, and whatever this water touched it contaminated. As a result, every house in the flooded district was filled, in most cases to the second floor, with most offensive matter. There was not a place which the flood touched where a man could lay his head with safety."

One of the biggest and most unpleasant jobs was digging out hundreds of cellars. Where houses had been swept from their foundations, it was often close to impossible to tell where they had been, the cellars having been so completely filled in by the flood. But every one of them had to be shoveled clean and their foul contents hauled out of the city.

The work camps themselves were another part of the problem,

with their "open plumbing, openly arrived at," as one doctor described it. Within the first week there were something like ninety-eight doctors at work, who, along with the Army, helped enforce basic sanitary practices. But what helped most of all was the weather. For ten days following the flood, temperatures stayed low, skies clouded, and there were frequent drenching rains. It was, wrote Dr. Lee, head of the sanitation board, a "great advantage in delaying decomposition." It was also miserable weather to work in or to be camping in amid the reeking muck.

On Friday morning Captain Bill Jones was back home in Pittsburgh, close to a state of collapse after working four straight days with almost no sleep or rest. Jones had paid out of his own pocket for the supplies he had brought to Johnstown and the wages of the men he had taken with him. In an interview with the Pittsburgh *Press* he gave great credit to the leadership Moxham had shown and added that he had had a hundred or so Hungarians working for him and that they had "worked like heroes."

General Hastings held up handsomely against the strain. He rode about town on a big horse, waving his floppy hat, and, in the main, did a good job. Director James Scott stayed on day after day, working with boundless energy. He wound up growing a full beard during his time in Johnstown and lost thirty pounds.

But perhaps the most resilient worker of them all, and certainly the one who stirred up the most talk, was a stiff-spined little spinster in a plain black dress and muddy boots who had brought the newly organized American Red Cross in from Washington. Miss Clara Barton and her delegation of fifty doctors and nurses had arrived on the B & O early Wednesday morning.

Clara was sixty-seven. She had been through the Civil War, the Franco-Prussian War, and several nervous breakdowns. For a while she had tried running a women's prison in Massachusetts. But since 1881, when, after a long campaign, she had succeeded in establishing an American branch of the International Red Cross, little else had been of real interest to her. And though her position as president was only a part-time job, she had already been down the Ohio by river barge to help during the floods of 1884 and to Texas with food and supplies during the famine of 1887. She had taken her workers to Illinois after a tornado in '88, and later that same year to Florida during a yellow-fever epidemic.

But these had been minor challenges compared to Johnstown, which she realized the moment she saw the valley from her train window. When the news of the disaster had first reached Washington Friday night, she had postponed doing anything for twenty-four hours; the story seemed too frightful and improbable to be true. But once there she knew that her Red Cross had arrived at its first major disaster. The organization, she had long argued, was meant for just such emergencies, and now she intended to prove it.

She set up headquarters inside an abandoned railroad car and, using a packing box for a desk, began issuing orders. Hospital tents were to be opened immediately, construction was to start on temporary "hotels" for the homeless, and a house-to-house survey was to be conducted to see just how many people needed attention. Hastings, it appears, did not know quite what to make of all this, or of Clara. "I could not have puzzled General Hastings more if I had addressed him in Chinese," she wrote later, adding that "the gallant soldier could not have been more courteous and kind."

In her stocking feet Clara stood five feet tall. She had a prominent nose, bright, black eyes, and a resolute set to her mouth. She took what little sleep she needed on a hard, narrow cot and had no use for demon rum, bumbling male officials, or for that matter anyone who attempted to tell her how to run her business. "A keen, steadfast, powerful New England woman," she was described as by one writer.

Within a very short time several large tents were serving as the cleanest, best-organized hospitals in town; six Red Cross hotels, two stories tall, with hot and cold running water, kitchens, and laundries, had been built with some of the fresh lumber on hand; and Clara herself was situated in her own command tent with the Red Cross banner flying overhead. When the survey was completed it was found that a large number of people with serious injuries had been too weak or broken in spirit to do anything to help themselves. Moreover, a surprising number of cases of prolonged shock were discovered, a phenomenon that would also be noted by a correspondent for the *Medical News* of Philadelphia. "A profound melancholia," he called it, "associated with an almost absolute disregard of the future" and evidenced by "a peculiar intonation of words, the persons speaking mechanically."

Clara and her people did their best to tend everyone they could. Clara herself worked almost round the clock, directing hundreds of volunteers, distributing nearly half a million dollars' worth of blankets, clothing, food, and cash. She also spoke her mind once or twice to the head of the Philadelphia chapter of the Red Cross, with the result that within a few days neither group would have anything to do with the other.

There seems little doubt that except for Hastings she was the best-known, among the people of Johnstown, of all the outsiders on hand and certainly the one who would be remembered the longest. She stoutly proclaimed that the Red Cross was there to stay as long as there was work to do. "We are always the last to leave the field," she said. She seemed to be everywhere at once, bouncing through the streets in a buckboard, scrutinizing the way things were being handled, whether she had anything to do with them or not. On one such ride she was accompanied by an Episcopal priest who was afraid she would be jolted to pieces and told her so. "Oh, this is nothing," she shouted back, "so long as we have no bullets flying around us."

Clara stayed for five months, never once leaving the scene even for a day. In October, when she did finally pack up and go, it was with all sorts of official blessings and thanks. She was presented with a diamond locket from the people of Johnstown; glowing editorials were written (". . . too much cannot be said in praise of this lady . . . To her timely and heroic work, more than to that of any other human being, are the people of the Conemaugh Valley indebted . . ."); and back home in Washington a dinner was given in her honor at the Willard with the President and Mrs. Harrison in attendance. The Red Cross had clearly arrived.

There were others who would be remembered. A few small-time crooks slipped in with the crowds. They queued up with the flood victims to collect whatever the Red Cross happened to be handing out at the moment or grabbed what they could from the debris. One or two suspicious-looking characters were nabbed before they had a chance to do much of anything and were swiftly hustled out of town. There were some, too, who simply hung around long enough to educate themselves on the place, then lit off to play on the sympathy of neighboring towns, describing the horrors of the devastation and their own heart-rending experiences in the flood at one back doorstep after another. One of them who went straight to Pittsburgh even managed to take the Relief Committee there for a good deal of cash before he was found out.

A handful of crackpots appeared on the scene, most of whom were of the religious-fanatic sort, and the most memorable of them was a gaunt prophet from Pittsburgh known as "Lewis, the Light," who wore nothing but long, red underwear and passed out handbills that said, among other things:

> Death is man's last and only Enemy, Extinction of Death is his only hope. Your soul, your breath, ends by death. Whew! Whoop! We're all in the soup. Who's all right? Lewis, the Light.

Then there were the "harpies," as some of the newspaper correspondents called them, who apparently came in from Pittsburgh. It was reported that a number of them had been seen at points along the line from Johnstown to Pittsburgh trying to recruit girls for "their nefarious calling" among the prettiest flood victims. Several who had tried to board the trains had been put off, but enough of them managed one way or other to reach Johnstown, so that within not too long a time Prospect Hill was making its own lively contribution to the mining-camp spirit of the place. Prospect "is overridden with bawdy houses and places for the illicit sale of liquor," one police official said.

This in turn added fuel to the cause of the W.C.T.U. ladies who had arrived almost immediately after the disaster and stayed on as the valley filled up with more and more men and as the demand for strong drink grew apace. Through most of June, Hastings kept the lid on liquor sales; the only thing being served was lemonade, which was sold at little makeshift stands through town. When, in July, the liquor ban was lifted, the lemonade concessions went immediately out of business. Twenty-two celebrants were arrested the first day, and George Swank wrote in the *Tribune* that the lemonade had "made more sickness than all the beer and whiskey that could be drunk."

And, of course, there were the sight-seers. They had been coming almost from the first morning after the disaster. They came into South Fork by special trains from Altoona. They arrived in

Johnstown on excursion trains that chugged in along the B & O weekend mornings. On Sunday, June 23, several hundred arrived, turned out in holiday attire and carrying picnic baskets. It seemed incredible; but there they were. The wreckage at the stone bridge seemed to fascinate them the most. But they strolled about everywhere, got in the way, set up their lunch parties inside abandoned houses, laughed, took pictures, asked a lot of silly questions, and infuriated nearly everyone except a few enterprising local men who set up booths and began selling official Johnstown Flood relics: broken china, piano keys, beer bottles, horseshoes, buttons, even bits of brick or wood shingles.

Hastings and the other officials kept asking the railroads to stop selling tickets to anyone who had no rightful business in Johnstown; before the month was out the railroads had agreed to comply as best they could and the number of visitors dropped off sharply.

Actually there would have been many more sight-seers than there were had there not been such widespread fear of disease; and though the cooperation of the railroads helped considerably to diminish the problem, the real turn came when it became known that typhoid fever had actually broken out in the valley.

The first clear-cut case was identified on Monday, June 10. More cases were found in the next several days, but the news was kept quiet. Within a week, however, the disease had spread swiftly and so had the rumors. By July 25 there would be 215 cases of typhoid within the flooded area and 246 beyond.

The doctors and sanitation crews, already dog-tired, flew into a frenzy of activity, working night and day to stop the thing and to keep the valley from panicking. Considering what might have happened under the circumstances, they did a spectacular job. But before the summer ended, forty people would die of typhoid; and like those who died of injuries or exposure in the first days after the disaster, they would not be included in the official figure given for the flood's victims.

Interestingly enough, along with the typhoid, there ran a long spell of unusually good health in Johnstown. There were fewer colds, fewer cases of measles, and fewer people complaining of "spring disorders" than there would have been under normal conditions. Those who managed to stay healthy through the first fright-

ful days immediately following the disaster, stayed very healthy
from then on. The regulations and precautions enforced by the sanitation people undoubtedly had a great deal to do with this, but so,
most everyone would later agree, did the fact that there was by the
end of the first week something almost like a spirit of exhilaration in
the air. There was so much happening all around. Every day there
was some kind of excitement. Those who had survived, despite how
much they may have suffered, began to discover new energy in
themselves. They were alive, and bad as things were that was still a
lot better than being dead. And there was now so very much to be
done.

There were exceptions, to be sure. Dr. Swan, for example, was
so broken in spirit by what he had been through that he would
never be able to practice medicine again. And the only fatality
among the polyglot army of people brought in to help was a suicide
among General Hasting's troops. Sunday afternoon a moody farm
boy with a wife and two children back home became so depressed
by what he had seen that he went into his tent and shot himself
through the head.

But for nearly everyone else the almost absurd idea that they
were going to pick up and start over again, to rebuild everything,
began working like a tonic. They started pulling together what was
left of their old lives and got to work on the new.

Most of them had precious little left of the old. Dr. Matthews,
who had opened a new practice in Johnstown just a few months
before the flood, found not a shred of all that he had owned except
for one shaving mug. Of George and Mathilde Heiser's earthly possessions, the only thing recovered was a big wardrobe which young
Victor retrieved from the wreckage on Main Street. When he pried
open the door, he found his father's old Civil War uniform inside,
and in one pocket a single penny. It represented his entire inherit-
ance.

Men like Horace Rose and John Fulton had lost their homes
and virtually everything else they owned. James Quinn and his
brother-in-law were able to recover scarcely a single board from
their dry-goods store. The only possessions recovered from the
Quinn home were some books, a photograph album, and some silverware, which were found scattered a mile or more from where
the house had stood.

Quinn, like hundreds of others, had sent his children off to Pittsburgh to live with relatives, while he stayed on to do what he could to help. But there were a few days before they left which the children would remember always. For nearly every child who survived, the week after the flood was a time of high adventure. The dynamite blasting at the bridge, the commissaries with their wondrous stacks of goods, the mobs of strangers tramping through the streets, the Iron Company's little wood-burning switching engines with their bell-shaped stacks shunting back and forth moving supplies, the nuns, the soldiers, the Pittsburgh firemen in their long rubber coats, were all something to see.

Gangs of small boys made great sport of sneaking past the sentries posted about town, or crawling among the rows of tents. Little prefabricated houses, called "Oklahomas," were being put up everywhere, and often, if he acted sensible enough, a boy could get a chance to help out. And every child, it seemed, took up relic hunting. "Every yard would yield something if one had the energy to dig," Gertrude Quinn wrote later. Gold and jewelry were supposedly being found all over town. One story had it that a crockery jar full of $6,500 worth of gold had been found in the mud where old William Macpherson had had his grocery store. Gertrude Quinn's sister Marie found a two-and-a-half-dollar goldpiece. And it was no trick at all to find half dimes or three-cent pieces or even a shirt stud with what looked like a diamond in it.

On Sunday, the 9th, the sun broke through for the first time since the flood. The spring green of the hills gleamed in bright morning sunlight, and overhead there were only a few small, soft clouds, and all the rest of the sky was a clean blue. Work went on the same as any other day. The air, already warmer than it had been for weeks, rang with the sound of picks and axes, hundreds of hammers, and, strangely it seemed at first, church bells.

On the embankment near the depot, just back from where Hastings had his headquarters, the chaplain of the 14th Regiment, H. L. Chapman, and David Beale began conducting the first services since the flood.

There were no more than thirty people gathered at first, but as time passed the crowd grew. Soldiers hanging about nearby drifted over. People came from the depot and over from the center of town. Beale stood on a packing box and told a story about overhearing a newcomer in the valley ask a small boy how bad things were in Johnstown, to which the boy was said to have replied, "If I was the biggest liar on the face of the earth I could not tell you half."

The rest of the service went as might be expected until John Fulton got up and started saying things that soon had the crowd stirring.

He said the Cambria shops would be rebuilt. "Amen!" answered several voices. "Johnstown is going to be rebuilt," he said. "Thank God!" someone answered again.

He said he could not speak for the Gautier works, but he was sure, nonetheless, that they, too, would be rebuilt, and bigger than ever. The Cambria men would be taken care of, he told them, and if you still have your family left, he said, then "God bless your soul, man, you're rich." His sermon was simple: "Get to work, clean up your department, set your lathes going again. The furnaces are all right, the steel works are all right. Get to work, I say. That's the way to look at this sort of thing . . . Think how much worse it could have been. Give thanks for that great stone bridge that saved hundreds of lives. Give thanks that it did not come in the night. Trust in God. Johnstown had its day of woe and ruin. It will have its day of renewed prosperity. Labor, energy, and capital, by God's grace, shall make the city more thriving than ever in the past."

"Amen!" again from the crowd.

That Johnstown should be rebuilt was by now taken as a matter of course. That it should be rebuilt right where it was before was also something everybody took for granted. No matter how dreadfully the valley had been ravaged, it was still their home. In fact, there is no record of anyone ever seriously considering the idea of not rebuilding in that particular place. The only question now was how long was it going to take to get things rolling once more.

The scene on the hillside would be remembered for years. Fulton, tall and spare, with his iron-gray beard and dark brow, certainly looked and sounded like a man of God; but he was also, as everyone knew, the voice of the Cambria Iron Company speaking, and as sincere as he may have been in asking them to trust in the Lord, his audience had had somewhat more experience putting its trust in the Cambria Iron Company. And either way, any man who could speak for both God and the Iron Company was someone to listen to closely.

The idea that the stone bridge had actually saved lives was a new one to most people, but the more they thought about it, the more they accepted it. The fire at the bridge seemed to epitomize the worst of the flood's horrors, but the fact was the death toll would have been far greater, perhaps even twice as great, had the bridge collapsed. The whole town would have been plunged down the valley to almost total destruction.

But it was when Fulton began talking of still another matter that a tense stillness came over the crowd and his every word could be heard over the sounds from the city below.

"I hold in my possession today," he said, "and I thank God that I do, my own report made years ago, in which I told these people, who, for purposes for which I will not mention, desired to seclude themselves in the mountains, that their dam was dangerous. I told them that the dam would break sometime and cause just such a disaster as this."

Then he changed to another subject. But he had said enough. In just two sentences he had hit on something that had been smoldering in people's minds for days. There had been plenty of talk about what had happened at South Fork and about the club. People were bitter, and with their renewed energy had come anger, deep and highly inflammable, and perhaps even contributing to that energy.

It was not that Fulton had been the first to raise the question of the dam and who was to blame for the flood; what was important was that he, John Fulton, had said what he had. The issue was now right out in the open, full-scale and officially. Moreover, there was no longer any question about which side the Iron Company might be on; and, perhaps most significant of all, if things came to a showdown, which everyone felt sure they would, Fulton, it would appear, held a piece of paper of considerable importance.

IX

"Our misery is the work of man."

———

The excitement in Pittsburgh continued day after day. Johnstown seemed to be the only thing people were talking about, and the papers carried almost nothing but flood news, with stories running on and on, page after page, and in even greater detail than what was being published elsewhere. The two cities had always had ties, through the steel business and family connections. Now almost everyone in Johnstown, it appeared, had relatives in Pittsburgh.

Refugees from the disaster kept pouring into Union Station by the thousands. The sick and the injured had to be cared for. Children, hundreds of them, lost or orphaned, many wearing tags for identification, had to be fed and looked after. Homes had to be found for them and all the others. On Wednesday, the 5th, four trains full of survivors, most of them women and children, came in.

People had dropped everything to help. Ladies' groups were sorting clothes and packing medical supplies in church basements all over town. The Masons, the Republican Club, factory workers were organizing, collecting, donating, and proudly announcing their accomplishments to the papers. The involvement grew so that the local merchants began complaining of a serious drop off in trade. "From a business point of view" things were the worst they had been for years, according to one report. Several firms canceled their regular newspaper advertisements in order to express their sympathies for the people of Johnstown, and Young's picture store on Wood Street attracted considerable attention by displaying in its window a painting of South Fork dam done a few years earlier by a local artist.

Everybody, it seemed, had his own latest story from Johnstown. A husband had heard from another man at the mill, a brother had just come back from the railroad depot, a cousin who worked at the Mellon bank had overheard something, two sons from the big Italian family across the alley had actually been there with one of the Flinn gangs, and they all had stories to tell, inside information. The city was alive with the most hair-raising tales and rumors. And nowhere was there more talk, or were things in such turmoil, than at the Pennsylvania depot and yards, where tons of food and supplies were still piling up, and the crowds were so thick, any hour of the day, that you could barely make your way through.

The railroad itself had never known such times, not even during the worst of the war years. Every schedule had been canceled. All normal business had been stopped. Nothing went east but trains bound for Johnstown, and as it was, the traffic was almost more than could be handled. If there had been only the storm damage to contend with, troubles would have been bad enough; but train after train kept steaming in from across the country, men and supplies had to be kept moving, repair equipment had to be sent forward, and everything that went up had to come back by the same route.

Pitcairn, with full authorization from the main office in Philadelphia, did everything possible to speed things up. The Pennsylvania had already donated $5,000 to the relief fund, but that was of small consequence compared to what was accomplished to keep the line open. Pitcairn himself worked almost without letup. All available manpower east and west was rushed into the area, and the cost of everything was assumed by the line.

This was by far the biggest emergency the Pennsylvania had ever been called on to face, and all its extraordinary power, its almost military-style discipline and organization, its vast resources in men and equipment, were brought to bear on the problem. The results, the swiftness and efficiency with which forces were marshaled, tangles unsnarled, damages repaired, help rushed through, were indeed remarkable and left a lasting impression on everyone involved. For all its highhanded ways, for all the evils people attributed to it, in a crisis the railroad had been worth more than any other organization, including the state, and they would remember that.

Still it would be two full weeks until the line from Harrisburg west would be opened and relief trains could start to Johnstown by way of Altoona. Until then Pittsburgh would remain the one channel through which everything had to pass.

The Allegheny River, with its endless freight of wreckage, also continued to be an immense fascination. Children were brought from miles away to watch the tawny water slip past the shores, so that one day they might be able to say they had seen something of the Johnstown Flood. The most disreputable-looking souvenirs, an old shoe, the side of a packing box with the lettering on it still visible, were fished out, dripping and slimy, to be carried proudly home.

There were accounts of the most unexpected finds, including live animals. But the best of them was the story of a blonde baby found at Verona, a tiny river town about ten miles up the Allegheny from Pittsburgh. According to the Pittsburgh *Press*, the baby was found floating along in its cradle, having traveled almost eighty miles from Johnstown without suffering even a bruise. Also, oddly enough, the baby was found by a John Fletcher who happened to own and operate a combination wax museum, candy stand, and gift shop at Verona.

Fletcher announced his amazing discovery and the fact that the baby had a small birthmark near its neck. Then he hired a pretty nineteen-year-old, dressed her in a gleaming white nurse's uniform, and put her and the baby in the front window of his establishment. Within a few days several thousand people had trooped by to look at the Johnstown baby and, it is to be assumed, to make a few small purchases from the smiling Mr. Fletcher. Then, apparently, quite unexpectedly, the baby was no longer available for viewing. The mother, according to Fletcher, had lived through the

flood and, having heard the story back in Johnstown, rushed to
Verona, identified the birthmark, and went home with her baby.

But there was another subject that was stirring up far more
talk in Pittsburgh. It was not until three or four days after the flood
that the rest of the country began growing keenly interested in the
South Fork Fishing and Hunting Club, but in Pittsburgh, not surprisingly, the interest had been high since news of the disaster first
came through Friday night. And for those who may have forgotten, or who never knew, the nature of the club's membership, the
Pittsburgh newspapers were quick to remind them.

In the beginning there had been some concern over those
clubmen who had been at the dam when it failed. But when it became known that they were alive and unharmed, the emphasis immediately shifted to what exactly the other club members might do
next.

On Saturday, at the mass meeting called by Pitcairn, Frick and
Phipps had been named to serve on the executive board of the Relief Committee. That night, at the home of another member,
Charles Clarke, a number of the clubmen met in private to agree on
what their policy should be. At that point, like everyone else in
Pittsburgh, they knew very little about what precisely had happened at the dam; but judging from the way things looked, the wis-
est policy for the moment seemed to be to say nothing, except that
no immediate action was planned and that the club would make a
donation to the people of Johnstown of 1,000 blankets.

But, unfortunately for the others, a few members decided to
speak their minds all the same. One member, who asked that his
name be withheld, told reporters that in the past he had heard questioning about the strength of the dam, but that he had never looked
into the matter personally. Then he told a story of riding from the
lake down to Johnstown a few years earlier with a driver who had
said, "The time will come when more than you and I will talk
about that embankment." And he finished up by saying that there
were some in Johnstown who used an Episcopal prayer, "Lord de-
liver us from mountain floods!"

Another member, James McGregor, who gave his name without any hesitation, said he refused to believe that there had been any
trouble at South Fork. He was certain the whole thing was a mistake.

"I am going up there to fish the latter part of this month," he said. "I am a member of the South Fork Fishing Club and I believe it is standing there the same as it ever was.

"As for the idea of the dam ever being condemned, it is nonsense. We have been putting in from twenty thousand to fifteen thousand dollars a year at South Fork. We have all been shaking hands with ourselves for some years on being pretty clever businessmen, and we should not be likely to drop that much money in a place that we thought unsafe. No sir, the dam is just as safe as it ever was, and any other reports are simply wild notions."

His own notions, which appeared in the papers on the morning of Sunday, June 2, were so wild, and so very tactless in the face of what was by then known of the suffering at Johnstown, that the only possible excuse for making such a statement must have been that he actually believed every word of it.

And to make matters worse, he was not alone. Young Louis Clarke next told a correspondent for the New York *Herald* that there was great doubt "among the engineers" who had examined the reservoir whether, after all, it had been that particular dam which broke. Just which engineers he was referring to is unclear, but he was interviewed along with another club member, James Reed, who said that in the past he himself had climbed all over the dam, studying it closely, and that "in the absence of any positive statement I will continue to doubt, as do many others familiar with the place, that it really let go." Perhaps, he then suggested, it had been a dam at Lilly which broke.

Reed's comments were of more than passing interest for he was the partner of Philander Knox in the prestigious Pittsburgh law firm of Knox & Reed. If there were to be lawsuits over the disaster, the South Fork Fishing and Hunting Club would almost certainly be represented by Knox & Reed. And already the press was playing up the likelihood that such suits would follow. On June 2 the *World* published a statement attributed to a prominent lawyer practicing in Allegheny County, who preferred to remain anonymous:

"I predict there will be legal suits with possible criminal indictments as a result of this catastrophe. I am told that the South Fork Club has been repeatedly warned of the unsafety of its dam, and it comes from good authority . . ."

On another page the *World* published an interview with Jesse

H. Lippincott of New York City, who was the son of a club member and who had spent several summers at the lake. The dam, he said, was built almost entirely of solid stone, but if it had indeed broken, the death toll would likely run to several thousand, and "Pittsburghers will . . . be deprived of their most popular resort."

Then, on Monday, the 3rd, reporters from Johnstown reached the dam and started sending a series of dispatches from South Fork which removed once and for all any fantasies about the dam still standing; and out of conversations with people in the neighborhood, they began building a history of the structure which did not bode well for the club members.

Feelings were running very strong against the club at South Fork. Monday after dark an angry crowd of men had gone up to the dam looking for any club members who might have been still hanging about. When they failed to find anyone, they broke into several of the cottages. Windows were smashed and a lot of furniture was destroyed. Then, apparently, they had gone over to the Unger farm to look up the Colonel. The reporters later called it a lynch mob and said they were bent on killing Unger. Whether or not it would have come to that, there is no way of knowing, for Unger by that time was on his way to Pittsburgh. There was a good deal of grumbling among the men as they milled about outside Unger's house; threats were shouted; then the men went straggling off through the night, back down the hollow.

The clubmen who had been at the lake had gone off on horseback, heading for Altoona, almost immediately after the dam broke Friday afternoon, though one of them, it seems, stuck around long enough to settle his debts with some of the local people. He had no intention of ever coming back again, he told them, which they in turn repeated for the benefit of the newspapermen. They also emphasized that the Pittsburgh people had not made things any better for themselves by pulling out so rapidly at a time when, as anyone could see, there was such a crying need for able-bodied men in the valley. Had they stayed on to help, it was said, then people might have felt somewhat differently toward them. This way there was only contempt.

But it was when they began describing how the dam had been rebuilt by Ruff and his workers that their real bitterness came through, that all the old, deep-seated resentment against the rich, city men began surfacing. Farmers recalled how they had sold Ruff hay to patch the leaks. A South Fork coal operator who insisted that his name be withheld, but who was almost certainly George Stineman, South Fork's leading citizen, told how, years earlier, he had gone to Johnstown on more than one occasion to complain about the dam's structural weaknesses. Reporters heard that the dam had been "the bogie of the district" and how it had been the custom to frighten disobedient children by telling them that the dam would break. The clubmen were described as rude and imperious in their dealings with the citizens of the valley. Reporters were told of the times neighborhood children had been chased from the grounds; and much was made of the hated fish guards across the spillway. Old feuds, personal grudges, memories of insults long forgotten until then, were trotted out one after the other for the benefit of the press.

Someone even went so far as to claim that several of the Italian workmen employed by the club had been out on the dam at the time it failed and had been swept to their death, thus implying that the Pittsburgh men had heartlessly (or stupidly) ordered them out there while they themselves had hung back on the hillsides.

One local man by the name of Burnett, who conducted a reporter on an inspection of the dam, told the reporter that if people were to hear that he was from Pittsburgh, they might jump to the conclusion that he was connected with the club and pull him from the carriage and beat him to death. "That is the feeling that predominates here," Burnett said, "and, we all believe, justly."

The plain fact was that no one who was interviewed had anything good to say about the South Fork Fishing and Hunting Club, its members, or its dam. And when a coroner's jury from Greensburg, in Westmoreland County, showed up soon after the reporters, the local people willingly repeated the same things all over again.

The jurymen had come to investigate the cause of death of the 121 bodies that had been recovered at Nineveh, which was just across the line in Westmoreland County. They poked about the ruins of the dam, talked to people, made notes, and went home. The formal investigation, with witnesses testifying under oath, was to be held on Wednesday, the 5th.

In the meantime, Mr. H. W. Brinkerhoff of *Engineering and Building Record*, a professional journal published in New York, arrived in South Fork to take a look at the dam and was soon joined by A. M. Wellington and F. B. Burt, editors of *Engineering News.* Most of the reporters remained cautious about passing judgment on the dam, waiting to see what the experts had to say. But on June 5 the headline on the front page of the New York *Sun* read:

CAUSE OF THE CALAMITY

The Pittsburgh Fishing Club
Chiefly Responsible

The Waste Gates Closed
When the Club Took
Possession

The indictment which followed, based on a *Sun* reporter's "personal investigation," could not have been much more bluntly worded.

> . . . There was no massive masonry, nor any tremendous exhibition of engineering skill in designing the structure or putting it up. There was no masonry at all in fact, nor any engineering worthy of the name. The dam was simply a gigantic heap of earth dumped across the course of a mountain stream between two low hills. . . .

In Johnstown on the same day, General Hastings told a *World* correspondent that in his view, "It was a piece of carelessness, I might say criminal negligence." In Greensburg the Westmoreland coroner's jury began listening to one witness after another testify to the shoddy way the dam had been rebuilt and the fear it had engendered, though two key witnesses had apparently had second thoughts about speaking their minds quite so publicly and refused to appear until forced to do so by the sheriff.

Two days later, on the 7th, a verdict was issued: ". . . death by violence due to the flood caused by the breaking of the dam of the South Fork Reservoir . . ." It seemed a comparatively mild statement, considering the talk there had been and coming as it did on the same day as Hastings' pronouncement. But on the preceding day, another cornoner's inquest, this one conducted by Cambria

County, had rendered a decision that spelled out the cause of the disaster, and fixed the blame, in no uncertain terms.

The Cambria jurors had also visited the dam and listened to dozens of witnesses. But their inquest was held to determine the death of just one flood victim, a Mrs. Ellen Hite. Their verdict was "death by drowning" and that the drowning was "caused by the breaking of the South Fork dam."

But then the following statement was added:

"We further find, from the testimony and what we saw on the ground that there was not sufficient water weir, nor was the dam constructed sufficiently strong nor of the proper material to withstand the overflow; and hence we find the owners of said dam were culpable in not making it as secure as it should have been, especially in view of the fact that a population of many thousands were in the valley below; and we hold that the owners are responsible for the fearful loss of life and property resulting from the breaking of the dam."

Now the story broke wide open. "THE CLUB IS GUILTY" ran the *World*'s headline on June 7. "Neglect Caused the Break . . . Shall the Officers of the Fishing Club Answer for the Terrible Results."

The Cincinnati *Enquirer* said that in Johnstown, as more facts became known, the excitement was reaching a "fever heat" and that "it would not do for any of the club members to visit the Conemaugh Valley just now." The Chicago *Herald* said there was "no question whatever" as to the fact that criminal negligence was involved.

Although it would be another week before the engineering journals would publish their reports on the dam, the gist of their editors' conclusions had by now leaked to the press. On Sunday, the 9th, *The New York Times* headline ran:

An Engineering Crime
The Dam of Inferior Construction
According to the Experts

Actually, the engineering journals never worded it quite that way. The full report which appeared in the issue of *Engineering News* dated June 15 said that the original dam had been "thoroughly well built," but that contrary to a number of previously published descriptions, it had not been constructed with a solid masonry core. (From this some newspapers would conclude that the "death-dealer" was nothing but a "mud-pile.") The repairs made by Benjamin Ruff, however, had been carried out "with slight care," according to the report. Most important of all, there had been "no careful ramming in watered layers, as in the first dam." But Ruff's work was not the real issue, according to the editors. "Negligence in the mere execution of the earthwork, however, if it existed, is of minor importance, since there is no doubt that it was not a primary cause of the disaster; at worst, it merely aggravated it."

The primary causes, it was then stated, were the lowering of the crest, the central sag in the crest, the fact that there were no outlet pipes at the base, and the obstruction of the spillway. The details of these matters were carefully described, and it was speculated that the disaster might have been averted that Friday afternoon if the bridge over the spillway and the fish guards had been cut away in time, or if some "man of great resolution, self-confidence, and self-sacrifice" had (as John Parke had contemplated) cut the dam at one end, where the original and more firmly built surface would have held up better against the enormous force of erosion.

But the point the editors of the report seemed most determined to hammer home was that there was no truth to any claims being made that the dam had been rebuilt by qualified engineers.

"In fact, our information is positive, direct, and unimpeachable that at *no time during the process of rebuilding the dam was* ANY ENGINEER WHATEVER, young or old, good or bad, known or unknown, engaged or consulted as to the work,—a fact which will be hailed by engineers everywhere with great satisfaction, as relieving them as a body from a heavy burden of suspicion and reproach."

Moreover, contrary to some statements made in Pittsburgh since the disaster, they had found no evidence that the dam had ever been "inspected" periodically, occasionally, or even once, by anyone "who, by any stretch of charity, could be regarded as an expert."

In other words, the job had been botched by amateurs. That

they had been very rich and powerful amateurs was not considered relevant by the engineering journals, but so far as the newspapers were concerned that was to be the very heart of the matter. It was great wealth which now stood condemned, not technology.

The club had been condemned by the coroners' juries, General Hastings, *and* by the engineering experts. The newspapers made no effort to investigate the dam themselves, and only one or two made any effort to present the facts about the dam or to explain even in passing why it had failed. Nor did the editorial writers make an effort to remain even moderately objective until more information became available. The club was guilty, criminally guilty several papers were saying, and that was that. Unlike the Hungarian stories, this one, it seemed, would hold up. It was based on about as solid information as could be hoped for, and in terms of its emotional content, it was perhaps even stronger. Now across the country there arose a great howl of righteous indignation.

For everyone who had been asking how such a calamity could possibly happen in the United States of America, there now appeared to be an answer, and it struck at the core of something which had been eating at people for some time, something most of them had as yet no name for, but something deeply disturbing.

For despite the progress being made everywhere, despite the growing prosperity and the prospect of an even more abundant future, there were in 1889 strong feelings that perhaps not all was right with the Republic. And if the poor Hungarians of Johnstown were signs of a time to come when a "hunky" could get a job quicker than a "real American," then the gentlemen of the South Fork Fishing and Hunting Club were signs of something else that was perhaps even worse. Was it not the likes of them that were bringing in the hunkies, buying legislatures, cutting wages, and getting a great deal richer than was right or good for any mortal man in a free, democratic country? Old-timers said that with every gain they made people were losing something. If that was so, people were beginning to think a little more about just what it was they might be losing, and to whom. And the more they thought about it, and especially the workingmen, the less they liked it.

It would be another three years before this kind of feeling would burst out in the terrible violence of the Homestead steel strike in Pittsburgh and Henry Clay Frick would nearly die of a bullet in his neck. And it would be another several years after that before public indignation over the power of the trusts, the giant corporations, and the men who ran them would erupt into public outrage. But the feeling was there in 1889, and it ran a great deal deeper than most people would have supposed. Certainly the language used by the press reflected a level of scorn and bitterness that would have been unthinkable a decade earlier.

The South Fork Fishing and Hunting Club was now described as "the most exclusive resort in America," and its members were referred to as millionaries, aristocrats, or nabobs. According to the Cincinnati *Enquirer* not even vast wealth was enough to gain admission, unless it was hereditary. "Millionaries who did not satisfy every member of the club might cry in vain for admission," the *Enquirer* wrote. "No amount of money could secure permission to stop overnight at the club's hotel . . ." The paper said that no one could visit the club without a permit, and called it "holy ground consecrated to pleasure by capital," but added that no one would want to go there now, "except to gaze a moment at *the Desolate Monument to the Selfishness of Man* . . ." J. J. McLaurin, the Harrisburg newspaperman, who was otherwise relatively reliable in his reporting on the disaster, wrote: "The club was excessively aristocratic, and so exclusive that Tuxedo itself might pronounce the Lorillard ideal a failure. The wealthy members never deigned to recognize the existence of the common clay of the neighborhood, farther than to warn intruders to keep off the premises."

Like dozens of others, McLaurin was also infuriated over the idea that the lake had served as a summer resort. He wrote that "50,000 lives in Pennsylvania were jeopardized for eight years that a club of rich pleasure-seekers might fish and sail and revel in luxurious ease during the heated term."

For an age which by no means looked upon pleasure as something to be expected in life, let alone life's chief objective, the very fact that the lake had been put there solely for *pleasure* seemed almost more than anyone could take; and in several editorials the writers seemed to imply that if the lake had served some other purpose, some *practical* purpose, then the tragedy would not have been quite so distressing.

"It is an aggravation of the calamity to reflect that the reservoir which gave way served no useful purpose, but merely ministered to the amusement of a gentleman's club composed of million-aires," wrote a small-town newspaper in New England. "The dam served no useful end, beyond the pleasure of a few rich men," said the *Daily Graphic* in New York. And the Chicago *Herald* pub-lished a cartoon showing what were supposedly seven clubmen done up in loud-checked coats and diamond stickpins, tossing down champagne on the clubhouse porch, while in the valley below them Johnstown is being wiped out.

Like several other papers, the *Herald* likened the clubmen to the Romans. "These wealthy sportsmen, these pleasure-seekers, sat in a secure place, in the amphitheater, like the noble Roman specta-tors when they gave the signal when the wild beasts were to be admitted into the arena to rend the bodies of the human victims. The Pittsburgh pagans did not give the signal, but they were just as guilty in the fact that they were told that the massacre was about to occur and made no effort to stop it . . ."

The effort alluded to here was the failure to remove the fish guards, which, very quickly, had come to symbolize everything re-pellent about the South Fork Fishing and Hunting Club. ". . . To preserve game for some Pittsburgh swells the lives of fifteen thousand were sacrificed," wrote the *Herald*. ". . . The ghosts of Johnstown are the ghosts of American labor that is dead." And a man by the name of Isaac Reed wrote a widely quoted poem which opened with the lines:

> Many thousand human lives—
> Butchered husbands, slaughtered wives,
>   Mangled daughters, bleeding sons,
>   Hosts of martyred little ones,
> (Worse than Herod's awful crime)
> Sent to heaven before their time;
>   Lovers burnt and sweethearts drowned,
>   Darlings lost but never found!
> All the horrors that hell could wish,
> Such was the price that was paid for—fish!

Interestingly, for all the abuse that was flung at the Pittsburgh people, very few newspapers ever went so far as to mention any specific names of members, and those that did mentioned only a half dozen or so. The Philadelphia *Press*, for all its superb coverage of what was going on in Johnstown, said hardly a word about the club or its members, and perhaps, as was hinted by another paper, because one club member, Calvin Wells, was a major stockholder in the *Press*.

As might be expected, the Pittsburgh papers were extremely cautious about printing anything untoward about the club, or, in some cases, were outright sympathetic toward the renowned members. The Pittsburgh *Press*, for example, took the position that too much scorn was being heaped on the club, since the dam had been built a long time back and the disaster, therefore, could as easily have happened at some earlier time. The *Post-Gazette* also felt the clubmen were being unfairly chastised. And Connelly and Jenks, authors of the so-called *Official History* of the flood, which was being written in Pittsburgh about that time, went out of their way to counteract popular images of opulent splendor at the lake. It was no center of pagan pleasure seeking or vulgar display, they wrote, but a place where the members of the club with their families and friends could "rough it" throughout the summer months. It was, they said, a comfortable, homelike place and as different from the "ordinary fashionable summer resort" as could be imagined. As for stories of any highhanded ways with the local people, well, "The place was exclusive only in the sense that a private house or garden is of that character. There was no lofty disregard of other people's rights, nor any desire on the part of the members to set themselves above those around them. The club was a happy family party, and nothing more."

Forest and Stream, a national fishing and hunting magazine, took strong objection to the "paragraphs hot with indignation" that were being published. Such stuff was easy to write, said the magazine's editors, who rose to the defense of the club largely on the grounds that its members were sportsmen who appreciated the beauties of the natural world and so, therefore, were essentially good men. Also, in the opinion of the editors, it was nonsense to condemn the clubmen because their lake was meant for pleasure. "To maintain a dam to form a lake for pleasure purposes is," they argued, "an enterprise no less legitimate than to build a dam for running a mill wheel." If the warnings about the stability of the dam had gone unheeded, perhaps that had been because the members were so preoccupied with the joys of life in the out of doors. And, concluded *Forest and Stream* there ought to be some compassion for the members, who in their hearts must surely be suffering terribly.

There were many, too, who looked upon the disaster as a time of the apocalypse. Countless sermons on "The Meaning of the Johnstown Flood" were delivered in every part of the land for many Sundays running. One Pittsburgh preacher compared the "wolf cry" about the dam breaking to those in his congregation who tired of hearing him on the admonitions of the Lord. Another said that the lesson was to be ever prepared to meet thy Maker.

In New York the illustrious Reverend T. DeWitt Talmage, using the 93rd Psalm as his text ("The floods have lifted up, O Lord, the floods have lifted up their voice; . . ."), told an audience of some 5,000 that what the voice of the flood had to say was that nature was merciless and that any sort of religious attitude toward nature meant emptiness. "There are those who tell us they want only the religion of sunshine, art, blue sky and beautiful grass," said Talmage. "The book of nature must be their book. Let me ask such persons what they make out of the floods in Pennsylvania."

Not a few ministers chose to talk about the spirit of sympathy that was sweeping the country. The New York *Witness,* a religious newspaper, went so far as to say there was a "loving purpose of God hidden in the Flood," which turned a great many stomachs in Johnstown.

But the theme that set the most heads nodding in agreement was the old, old theme of punishment from on high. The story of Noah was read from thousands of pulpits. ("And God looked upon the earth, and, behold, it was corrupt; . . . And God said unto Noah, The end of all flesh is come before me; . . .") This was The Great American Flood; it had been a sign unto all men, the preachers said, and woe unto the land if it were not heeded. The steel town had been a sin town and so the Lord had destroyed it; for surely only a vile and wicked place would have been visited by so hideous a calamity.

It was a line of reasoning which many people were quick to accept, for at least it made some sense of the disaster. But it was a line of reasoning which met with much amusement in Johnstown, where, as anyone who knew his way about could readily see, Lizzie

Thompson's house and several rival establishments on Green Hill had not only survived the disaster, but were going stronger than ever before. "If punishment was God's purpose," said one survivor, "He sure had bad aim."

There really was never much mystery in anyone's mind in Johnstown about the cause of the flood. George Swank spoke for just about everyone when he wrote, "We think we know what struck us, and it was not the hand of Providence. Our misery is the work of man."

The *Tribune* had started publishing again on the 15th. Swank referred to the Pittsburgh men as "the dudes" and said that they wanted "an exclusive resort where, in all their spotlessness and glory, they might idle away the summer days." The people of Johnstown, he said, had never had a chance. "A rat caught in a trap and placed in a bucket would not be more helpless than we were."

Dozens of Johnstown people spoke out against the dam, telling the out-of-town newspapermen what an awful menace it had been and describing the dread shadow of fear it had cast over their lives, and nearly every last one of them refused to give his name. The one outstanding exception was Cyrus Elder, Johnstown's only member of the South Fork Fishing and Hunting Club, who said that he had never considered the dam structurally faulty and, contrary to what John Fulton was saying, that he knew of no serious concern about the dam among the Cambria Iron people.

Having lost his wife and one daughter, his home and just about everything he owned but the clothes on his back, Elder had as much cause as anyone to lash out at the club, and certainly not to do so was to go against the temper of the entire town. But he stuck to his position. He admitted that Johnstown people had long been edgy about the dam and said, "Therefore, if anybody be to blame I suppose we ourselves are among them, for we have indeed been very careless in this most important matter and most of us have paid the penalty of our neglect." It was a brave and most unpopular thing to be saying in Johnstown. The statement was picked up immediately by the newspapers. But his line of reasoning was never given any serious consideration by the popular press, though *Engineering and Building Record* registered surprise that the men responsible for Johnstown's welfare, not to mention the officials of the Pennsylvania Railroad, with all that they had at stake, had not made sure that the lake over their heads was carefully built in the first place and properly maintained thereafter.

The railroad, for its part, remained quiet about any involvement it might have had in the dam's past, taking the position, no doubt, that its actions in bringing relief to Johnstown would speak a great deal louder and more favorably than any words—which indeed they did. And once the engineering journals had established that the so-called engineers from the railroad who, according to statements made by Pitcairn, had kept a watch on the dam were in no way qualified to make any sort of intelligent judgment, then there was really very little more that the railroad could say.

But if the club's guilt had been established as far as the newspapers were concerned, there still remained the matter of paying the penalty, and that such a penalty should be paid seemed self-evident.

One newspaper after another said that the club should have to make amends for what had happened. Not a little facetiously, *The New York Times* wrote, "Justice is inevitable even though the horror is attributable to men of wealth and station, and the majority of the victims the most downtrodden workers in any industry in the country."

Even the Boston *Post*, which except for the Pittsburgh papers was about as conciliatory toward the club as any paper, said that the members had better be prepared to pay up. The *Post*, quite generously, stressed that the members must have acted as most men would have under the circumstances, "trusting, perhaps not unjustifiably, to others" with no thought of imperiling the lives of anyone. "Even if all that is reported as to the construction of the dam proves true, there is the possibility that personally the owners were not guilty of the reckless parsimony attributed to them." Still, added the *Post*, "If they were unable or failed to cope with forces of nature which they called into action, the responsibility is theirs, and as they have sown so must they reap, even if the harvest is the whirlwind."

And behind every editorial was the suggestion of what the *Sun* said outright: "If they [the club members] should be held liable in civil suits for damages it is probable that many, if not all of them, will be financially ruined."

The Pittsburgh men had by now given some $6,000 to the relief fund, in addition to the 1,000 blankets, but that did not seem to help their cause much. "As they are almost all millionaires," wrote the New York *Daily Graphic,* "the sum is not staggering, but shows that, while they were negligent, they are not heartless. . . . Yet they should do more than they have for the sufferers. It was through their indifference that this great disaster was precipitated upon the residents of the peaceful valley. Remorse, if nothing else, should lead them to alleviate to the fullest extent of their wealth the suffering they have caused."

Very shortly thereafter several club members did, in fact, give generously; but, needless to say, it was far from the "fullest extent of their wealth."

Henry Clay Frick, through H. C. Frick Coke Company, gave $5,000. The Mellon family, through T. Mellon & Sons, gave $1,000. The Carnegie Company gave $10,000. There were several gifts of $1,000, $500, and $100. There was also one member who gave $15, and there were about thirty of them who never gave anything.

The members did suggest that the clubhouse could be used as a home for Johnstown orphans, but the offer was turned down with the excuse that the location was too inconvenient. There was also one member, S. S. Marvin, who actually went to Johnstown to see what he could do to help, and contrary to the many warnings published, he suffered no injuries, or even insults, from the people in the valley. Marvin had been appointed to one of the committees organized by the governor. He was in the baking business in Pittsburgh and had already contributed great quantities of bread. At Johnstown he looked about with absolute dismay and said, "Johnstown is a funeral," an expression the newsmen were quick to pick up.

As for the other members, they grew increasingly cautious about saying anything. Phipps, Mellon, and Knox said nothing at all. Unger, who was staying with his daughter in Pittsburgh, tried hard to play down the importance of the fish guards, saying that they were only a few feet high. He also reminded the reporters that the dam had been originally built by the state, thus implying that the matter of responsibility, if pursued, might become a very complicated piece of business.

Frick refused to see anyone from the press. Except for Carnegie, Frick was, of course, the best-known and most powerful of the

members, and unlike Carnegie, Frick had already had his name published in the papers as one of the members. Moreover, he was, after
Ruff, the ranking stockholder in the original organization and one
of the few founding members still in the organization. In other
words, he was one of the few people who had been involved in the
club at the time Ruff made his renovation of the dam. So anything
he might have to say would be of great interest, and possibly of
great importance to how things might go for the club in the courts.

But Frick was not talking, and it was probably not so much
that he was fearful of saying anything at that particular time as it
was that he simply did not talk to the press ever, at any time. It was
his standing policy. He was a highly uncommunicative sort anyway
and, by nature, abhorred all forms of notoriety. He had no trust in
newspapers, no liking for reporters, and talking to them, he was
convinced, was bad for business. Only once in his life did he break
his rule and speak freely to a reporter, but it was with the understanding that he could edit the copy, which he did, reducing a full
column to exactly ten lines.

In the weeks following the disaster Frick made no public statements, nor did he ever in later years.

Carnegie, on the other hand, had much to say, but never anything to suggest that he had had any connection with the club, and
almost no one was ever the wiser, since it would not be for another
year or more, when the story had been largely forgotten, that a
complete membership list was divulged. Carnegie was in Paris attending the World's Fair at the time the disaster occurred. When a
meeting of Americans had been called at the United States Lega-
tion, by the American Minister to France, Whitelaw Reid, it had
been Carnegie who put forth the resolutions quickly adopted by
the assembly. The people of Johnstown were to receive "profound
and heartfelt sympathy" from their brethren across the Atlantic;
they were also to be congratulated for their "numerous acts of
noble heroism" and especially were they to be admired for the way
they had "preserved order during chaos" through their own local
self-government. How much Carnegie then contributed to the
40,000-odd francs that were pledged is not known.

But as for the South Fork Fishing and Hunting Club and any
thoughts or feelings he may have had concerning its part in what
had happened, Carnegie made no mention of that, and there would

be none forthcoming. Carnegie wound up his affairs in Paris shortly thereafter, then left for his castle in Scotland, stopping off long enough in London to visit with the American Minister there, Robert Lincoln, the son of Abraham Lincoln.

Reporters in Pittsburgh, meanwhile, had been looking into the financial status of the South Fork sportsmen's association and had found, much to their dismay, that, for all the colossal wealth of the men who belonged to it, the club itself was capitalized for a mere $35,000 and there was a $20,000 mortgage still outstanding on the clubhouse. Since any future lawsuits would most likely be brought against the club, and not individual members, the chances for anyone collecting very much appeared to have diminished drastically. And just to be sure that no one missed this particular point, on June 12 James Reed once again granted the press an interview. Reed was a tall, sharp-faced man, quiet-spoken and scholarly looking. His practice included several of Pittsburgh's biggest concerns, as well as the Carnegie interests. His professional prestige was very high. What he had to say, therefore, was carefully taken down and later read with special interest.

The capital stock of the club would be the extent of the liability, he declared, if, that is, there were any liability, and in his opinion there was not. "I have tried," he said, "to divest myself of my identity with the South Fork Fishing Club to see if there could possibly be any grounds for a suit against the company or individual stockholders, and I am free to say I have been unable to find any. If a person was to come to me as an attorney and want me to bring suit against the company for damages resulting from the flood, I could not do so, because there are no grounds for such a suit."

Then, in conclusion he said, "As one of the stockholders I most certainly regret the sad occurrence, and I know the rest do; but I cannot see how the organization can be held legally responsible for the breaking of that dam."

But if he could not, there were others who could. At the end of July the first case brought against the South Fork Fishing and Hunting Club was filed at the Allegheny Court House in Pittsburgh, where the club had been originally incorporated. Mrs. Nancy Little and her eight children were suing the club for $50,-000 for the loss of her husband, John Little, a woodenware salesman from Sewickley, Pennsylvania, who had been killed at the

Hulbert House. The attorneys for the defense were, as had been expected, Knox & Reed, who filed a voluntary plea of not guilty. Then the case was put off for several months.

Early in August a group of Johnstown businessmen organized to sue the club. They raised some $1,300 to help meet expenses and hired John Linton and Horace Rose to start preparing their case.

Later on, James and Ann Jenkins, backed by some businessmen of Youngstown, Ohio, brought suit for $25,000 for the loss of Mrs. Jenkins' father, mother, and brother, who had been drowned at Johnstown.

There were also suits against the Pennsylvania Railroad, the most important of which was one filed in September by a Mr. Farney S. Tarbell of Pittsburgh. Tarbell accused the railroad of negligence in the death of his wife and three children, who had been passengers on the *Day Express*. There were suits for lost luggage, and a Philadelphia company sued for the loss of ten barrels of whiskey, which had been looted from a freight car. This last case was won by the Philadelphia company when a conductor admitted that he had looked the other way when the whiskey was being taken. It was, as things turned out, the only case won by any of those who brought suit against either the club or the railroad.

Not a nickel was ever collected through damage suits from the South Fork Fishing and Hunting Club or from any of its members. The Nancy Little case dragged on for several years, with the clubmen claiming that the disaster had been a "visitation of providence." The jury, it seems, agreed.

There is no account of how things went in court, as it was not the practice to record the proceedings of damage suits. Nor is there any record of the Jenkins case, though there, too, the clubmen were declared not guilty.

In the Tarbell case the judge acquitted the railroad, also designating the disaster a "providential visitation." And in Johnstown, after nearly two years of preparation, Colonel Linton and Horace Rose urged their clients to give up their suit, saying that it would almost certainly fail. The club had no assets, they argued, and there was no chance of winning unless individual negligence could be proved and that would be next to impossible since Ruff was dead. So Linton and Rose were paid $1,000 for their services and the suit was dropped.

Perhaps the most frustrating attempt to recover some retribution was carried on by Jacob Strayer, the Johnstown lumber dealer, who set out to sue the club for $80,000. The case sat for years, in one county court after another, as the club kept seeking a change of venue due to local feelings. Then after waiting something like five years without hearing anything, Strayer discovered that his lawyer, unbeknownst to him, had settled out of court (taking $500) and had died shortly after that. Strayer next went bankrupt; the club was long since insolvent; and nothing more happened.

Had the Little case or the Jenkins case been tried in Johnstown instead of Pittsburgh, it is possible that the decisions would have gone the other way, though in Johnstown there would have been small chance of finding twelve men to serve on a jury who would have been able to profess no bias against the club. In the judgment of lawyers who have examined the facts of the disaster in recent years, it also seems likely that had the damage cases been conducted according to today's standards the club and several of its members would have lost. It is even conceivable that some of those immense Pittsburgh fortunes would have been reduced to almost nothing. What the repercussions of that might have been is interesting to speculate. Possibly it would have delayed, perhaps even altered significantly, the nation's industrial growth.

In trying to evaluate why the cases went as they did, it is, of course, important to keep in mind the tremendous power of the people who were being sued. Their influence and prestige were such that few would have ever dared challenge them on anything. "It is almost impossible to imagine how those people were feared," Victor Heiser would say many years later. They were the ruling class. It was that simple. The papers could rail away to their heart's delight (while seldom ever mentioning any names), but to actually strike out at the likes of the clubmen, even within the confines of the courts, was something else again. Practically speaking, the odds against winning against them were enormous, even had the cases been open and shut, which they were not.

For to prove that any living member of the club had been personally negligent would have been extremely difficult. And in all fairness, it is quite likely, as the Boston *Post* suggested, that the clubmen themselves knew no more about the structural character of the dam than did anyone in Johnstown. Like nearly every leading citizen of Johnstown, with the exceptions of Morrell and Fulton, they made the mistake of assuming that the men who had rebuilt the dam had known what they were doing.

They had been told that the dam was properly engineered and properly maintained, and so, as long as everything went all right, they had no cause to think otherwise.

In addition, there is no doubt that the storm which brought on the failure of the dam was without precedent; or at least that during the relatively short period of time in which there had been some semblance of civilization in the area (which was less than a hundred years), no one had recorded a heavier downpour. So for such skillful lawyers as Knox and Reed to have argued that the whole dreadful occurrence was an act of God would have been very easy, and judging by the outcome, they made their point with great effect.

Certainly in the eyewitness testimony collected by the Pennsylvania Railroad in preparation for the suits it might have to face, repeated emphasis was placed on proving that no one had ever seen such a storm; and therefore if the "reasonable precautions" taken by railroad employees such as yardmaster Walkinshaw had turned out badly, it was only because the storm itself was so very unnatural. (It is also interesting to note that Pitcairn, in defense of Ruff's abilities, agreed openly that Ruff had no engineering training; Ruff was a lot better than any engineer, Pitcairn said.)

Still the heart of the matter remained the dam itself, and judging from occasional comments that appeared in the papers, it seems that the club's defense was based on the proposition that the dam would have broken anyway—even if it had had no structural flaws.

Apparently that was a convincing argument, despite the fact that several small dams which had been built near Johnstown to supply the city's drinking water had not failed as a result of the storm; and these, significantly enough, had been built under the personal supervision of Daniel J. Morrell.

The water in Lake Conemaugh, the attorneys for the defense must have claimed, was coming up so fast on the afternoon of the 31st, and would have continued to come up so fast, even had the dam held past 3:10, that eventually it would have started over the top, and once that happened, sooner or later, the best of earth dams would have failed. Even had there been no sag at the center, even if the spillway had been working to full capacity, the volume of water rushing into the lake was greater than what could get out, and so, they held, the end result would have been the same, except that it would have come later, and perhaps at night when the consequences would have been far more disastrous. It was a specious line of defense, for several reasons.

First of all, there is no way of ascertaining for certain whether the inflow of water was such that it would have caused the lake to spill over the breast of the dam for an extended period of time had the dam been higher at the center, instead of lower, and had there been no obstructions in the spillway. There is also no way of telling whether there was a drop off in the volume of water pouring into the lake in the hours following the break. In other words, would there have been enough water rushing off the mountain to keep the lake at a level higher than the breast of the dam (a properly engineered dam, that is) for many hours? It seems unlikely. Moreover, it was clear from the engineering studies made, and from photographs taken of the dam after the break, that it was that part of the dam which had been repaired by Ruff and his crew which went out on the afternoon of the 31st.

But even if it were assumed, for the sake of argument, that the Ruff repairs were as solid as the original dam, that the spillway obstructions did not greatly diminish its capacity, and that there was no sag at the center to reduce even further the spillway's usefulness, there still remains one very obvious and irrefutable flaw in the dam and in any argument in its defense.

Because there were no longer discharge pipes at the base of the dam, the owners never at any time had any control over the level of the lake. If the water began to rise over a period of days or weeks to a point where it was becoming dangerously high, there was simply nothing that could be done about it. If, on the other hand, the pipes had still been there, as they were up until they were removed by Congressman Reilly, or if new pipes had been installed by Ruff, then through that abnormally wet spring of 1889 the men in charge of the dam, Unger, John Parke, and others, could have kept the lake at a safe level of say at least ten to twelve feet below the crest of the dam.

So while there is no question that an "act of God" (the storm of the night of May 30–31) brought on the disaster, there is also no question that it was, in the last analysis, mortal man who was truly to blame. And if the men of the South Fork Fishing and Hunting Club, as well as the men of responsibility in Johnstown, had in retrospect looked dispassionately to themselves, and not to their stars, to find the fault, they would have seen that they had been party to two crucial mistakes.

In the first place, they had tampered drastically with the natural order of things and had done so badly. They had ravaged much of the mountain country's protective timber, which caused dangerous flash runoff following mountain storms; they obstructed and diminished the capacity of the rivers; and they had bungled the repair and maintenance of the dam. Perhaps worst of all they had failed—out of indifference mostly—to comprehend the possible consequences of what they were doing, and particularly what those consequences might be should nature happen to behave in anything but the normal fashion, which, of course, was exactly what was to be expected of nature. As one New England newspaper wrote: "The lesson of the Conemaugh Valley flood is that the catastrophes of Nature have to be regarded in the structures of man as well as its ordinary laws."

The dam was the most dramatic violation of the natural order, and so as far as a few rather hysterical editorial writers were concerned, the lesson of the flood was that dams in general were bad news. The writers took up the old line that if God had meant for there to be such things as dams, He would have built them Himself.

The point, of course, was not that dams, or any of man's efforts to alter or improve the world about him, were mistakes in themselves. The point was that if man, for any reason, drastically alters the natural order, setting in motion whole series of chain reactions, then he had better know what he is doing. In the case of the South Fork dam, the men in charge of rebuilding it, those who were supposed to be experts in such matters, had not been expert—either in their understanding of what they did or, equally important, in their understanding of the possible consequences of what they did.

What is more, the members of the club and most of Johnstown went along on the assumption that the people who were responsible for their safety were behaving responsibly. And this was the second great mistake.

The club people took it for granted that the men who rebuilt the dam—the men reputed to be expert in such matters—handled the job properly. They apparently never questioned the professed wisdom of the experts, nor bothered to look critically at what the experts were doing. It was a human enough error, even though anyone with a minimum of horse sense could, if he had taken a moment to think about it, have realized that an earth dam without any means for controlling the level of the water it contained was not a very good idea. The responsibility was in the hands of someone else, in short, and since that someone else appeared to be ever so much better qualified to make the necessary decisions and pass judgment, then why should not things be left to him?

In Johnstown most men's thoughts ran along the same general line, except that it was the clubmen who were looked upon as the responsible parties. And just as the clubmen were willing to accept on faith the word of those charged with the job of rebuilding the dam, so too were most Johnstown people willing to assume that the clubmen were dutifully looking to their responsibilities. If the dam was in the hands of such men as could build the mightiest industries on earth, who could so successfully and swiftly change the whole character of a city or even a country, then why should any man worry very much? Surely, those great and powerful men there on the mountain knew their business and were in control.

In the *North American Review*, in August 1889, in an article titled "The Lesson of Conemaugh," the director of the U. S. Geological Survey, Major John Wesley Powell, wrote that the dam had not been "properly related to the natural conditions" and concluded: "Modern industries are handling the forces of nature on a stupendous scale. . . . Woe to the people who trust these powers to the hands of fools."

It was, however, understandably difficult for the people of Johnstown ever to feel, like Cyrus Elder, that they too had been even partly to blame. Practically everyone felt that he had foreseen the coming catastrophe, and if he had not, like John Fulton, actually put anything down on paper, he, nonetheless, had been equally aware of the troubles with the South Fork dam and every bit as dubious about its future. That the members of the club were never required to pay for their mistakes infuriated nearly everyone in Johnstown and left a feeling of bitter resentment that would last for generations.

As for the club people, their summers at South Fork were over. The cottages sat high and dry along the vast mud flat which had been Lake Conemaugh and where, here and there, like the remains of some prehistoric age, stood the stumps of great trees that had been taken down more than fifty years earlier just before the dam had been built. By July grass had sprung up along South Fork Creek where it worked its way through the center of the old lake bed, and deer left tracks where they came down to drink.

For some time several cottages were occupied by Johnstown people. James McMillan, the plumber, and six or seven other men moved their families into the biggest of the houses, apparently with the consent of the owners, and, according to a notice in the *Tribune* at the end of July, the accommodations were as elegant as ever. But the owners themselves never came back, except for one, Colonel Unger, who not only returned, but lived out the rest of his life on the farm just above the remains of the dam. All the other property was broken up and sold off at a sheriff's auction.

In Johnstown the Cambria works had started up again by mid-July. It would be a long time before the furnaces were working to capacity, but nearly the full pay roll was being met and slowly things began to return to normal. Estimates were made on the total property damage (about $17 million). The banks opened. The Quicksteps were playing again. By August the Saturday night band concerts had been revived and a piano tuner had come to town. There was ice cream for sale; Haviland china was back on the shelves of those few stores that had been spared. A camera club was started, and the plumbers and steam fitters organized a union.

In September Mr. and Mrs. Andrew Carnegie came to town to see the flood damage, and before he left Carnegie had agreed to build a new library where the old one had been. In late fall the schools reopened.

New houses and shops were going up all over town. People who had fled the valley began coming back again. The Quinns were back by October, astonished to find how much had been accomplished and how bad the place still smelled. There was plenty of work still to be done, of course, and plenty of jobs, and it would stay that way for a long time. Just getting the place back to where it had been before would take five years or more.

But there were many who would leave Johnstown after the flood. For hundreds of people, like Victor Heiser, the disaster had deprived them of every meaningful connection with the place. Suddenly they were alone and there seemed no very strong reason for staying any longer, and particularly if they had ever had an ambition to see something of the world. Had his mother and father survived, Victor Heiser would have remained in Johnstown, probably, he would later speculate, to become a watchmaker. As it was, he left the valley within a year, and after working his way through medical school, spent most of his life as a public-health officer and physician, fighting disease around the world. He would also write a best-selling book on his experiences, *An American Doctor's Odyssey*, and would be credited with saving perhaps two million lives.

There were others who could stay no longer because of the memory of what had happened. And David Beale was among those who left out of bitterness over an experience during the frantic days which had followed the disaster. In Beale's case it had been a falling out with some of his congregation over the fact that he had turned the church into a morgue without authorization from the elders. There were plenty who rose to his defense, saying it had been the only intelligent and Christian thing to do under the circumstances and that somebody had to make the decision, but there were enough hard words exchanged to send Beale on his way to another charge in another town.

For years, too, there would be much speculation on how many of those people listed among the unfound dead were actually very much alive in some faraway place. It seemed reasonable enough to figure that some men, suddenly, in the first dim light of that terrible morning of June 1, had decided that here was an opportune time to quietly slip away to a new and better life. And if one of those names on the unknown list was somebody you had been close to, it was a whole lot pleasanter to think of him living on an apple farm in Oregon or tending bar in a San Francisco saloon than rotting away beneath six feet of river muck somewhere below Bolivar. Furthermore, such speculation seemed well justified when, eleven years later, in the summer of 1900, a man by the name of Leroy Temple showed up in town to confess that he had not died in the flood but had been living quite happily ever since in Beverly, Massachusetts. On the morning of June 1 he had crawled out of the wreckage at the bridge, looked around at what was left of Johnstown, then just turned on his heels and walked right out of the valley.

Stories of the flood would live on for years, and in time they would take on more the flavor of legends, passed along from generation to generation. Each family had its tales of where they had been when the wall of water came, where they ran to, who shouted what to whom, who picked up the baby, who went back for the horse, or how they had survived the night. Children who were only four or five years old at the time would live to be old men and women who would describe in the most remarkable detail how they had watched the flood strike the city (from a place where it would have been impossible to have seen the water) or how they had looked at their wrist watch at that exact moment (there were no wrist watches in 1889) and read (at age five!) that it was exactly such-and-such time. There would be stories of how grandfather tried to save an ax handle ("of all things!") or how Uncle Otto had thrown away his Bible when he saw what had happened. There would also be a great amount of durable gossip and some rather bad feelings about "certain people" who had somehow gotten their hands on more than their share of the relief money and how "their families are rich to this day because of it." And at least one Irish undertaker from Pittsburgh was said to have made "a positive fortune" out of the disaster.

There would also, one day, be signs posted in saloons from one end of the country to the other saying: PLEASE DON'T SPIT ON THE FLOOR, REMEMBER THE JOHNSTOWN FLOOD. At Coney Island and in Atlantic City re-creations of the great disaster would be major attractions for many seasons. And "Run for the hills, the dam has busted" would be a standard comedy line the country over for years.

In Johnstown three babies born on the fateful day would grow up with the names Moses Williams, Flood C. Raymond, and Flood S. Rhodes.

General Hastings would later be elected governor largely because of the name he had made for himself at Johnstown; and when William Flinn later became the Republican boss of Pittsburgh and a state senator, he made it a practice to remind election-year audiences of the job he had done at Johnstown.

Tom L. Johnson, who later gave up a lucrative business career to become the highly progressive (some said "socialistic") mayor of Cleveland, would use the flood to make a case for his political philosophy. In his autobiography he would write at length about the disaster and its cause and how charity had vitiated local energies (he was still Moxham's man in this regard). The flood, he would conclude, was caused "by Special Privilege," and: "The need of charity is always the result of the evils produced by man's greed."

In after-dinner speeches at the Duquesne Club, Robert Pitcairn would recall the services rendered by the railroad and ask if a "heartless corporation" could have behaved so. Bill Jones never said much about what he did, though he was quoted as saying that perhaps Johnstown ought to rebuild on higher ground. When he returned to Pittsburgh from Johnstown, Jones had only two months to live. At the end of the summer he was killed when a furnace he was working on at the Braddock mill exploded.

The members of the South Fork Fishing and Hunting Club remained silent. The matter of their part in the flood was simply not mentioned, and as the years passed less and less was heard of it. In another generation it would be just about forgotten in Pittsburgh.

The more or less agreed-to attitude of Johnstown's business people was also that the flood should be forgotten as soon as possible. There was no sense dwelling on the thing. It was bad for the spirits, and it most certainly was harmful to business.

"It may be well to consider that the flood, with all its train of horrors, is behind us, and that we have hence forth to do with the future alone," said George Swank in the *Tribune* on the morning of June 1, 1892. It was his conclusion to a long description of the ceremonies held the day before at Grandview Cemetery. The whole city had been shut down and close to 10,000 people had gone up to the new burying grounds.

Except for the plot for the flood dead, Grandview was still very sparsely occupied. It had been started by Cyrus Elder, John Fulton, and others only a few years before the flood and was laid out a good distance from town up on some of the highest land for miles around. The idea was that here the dead would be safe from spring floods. The view was very grand indeed, stretching off in every direction as far as the eye could carry; but the trees blocked a direct look back down into the great amphitheater among the hills where Johnstown lay, and so the city was wholly concealed, and except for the distant sound from the mills, it was almost as though there was nothing even like a city anywhere near.

On the afternoon of the 31st, with the new governor present, and with Johnstown's first mayor, Horace Rose, officiating, a large granite monument was dedicated to the "Unknown Dead Who Perished in the Flood at Johnstown, May 31, 1889."

Behind the monument, arranged very precisely row on row, were 777 small, white marble headstones.

The unknown plot had been purchased by the Relief Commission and the bodies moved there from Nineveh, Prospect Hill, and half a dozen other places during the early fall of 1889. It had taken the time since to raise money for the monument and the nameless headstones. Actually, there were not quite a full 777 bodies buried in the plot; someone had decided to set out a few extra stones just to make an even pattern. But the effect on the immense throng gathered in the warm afternoon sunshine was very great. Against the long sweep of grass and the darker green of the bordering trees, the people stood in their funeral best, clustered in a dark, tight mass, strangely motionless and silent beneath the veiled monument. A few steps beyond, the carriages for the dignitaries were drawn up.

There were several speeches, the longest and best of which was by the governor, Robert Pattison, who, in offering a lesson to be learned from the disaster, said, "We who have to do with the concentrated forces of nature, the powers of air, electricity, water, steam, by careful forethought must leave nothing undone for the preservation and protection of the lives of our brother men."

Then the choir sang "God Moves in a Mysterious Way"; the monument was unveiled, and people started back along the winding road that led down into town.

List of Victims

Total Number Lost, 2,209

This list, dated July 31, 1890, is the one printed in the Johnstown *Tribune* fourteen months after the Flood.

GRAND VIEW CEMETERY.

[Buried in private lots in Grand View.]

Alexander, Arailia K., Broad street.
Andrews, John, Sr., 57, John street.
Arther, Mrs. Alice, 29, Water street.
Bantley, William G., 36, Third Ward.
Bantley, Mrs. Ella, 30, Third Ward.
Bantley, George L., 6 months, Third Ward.
Barbour, Mrs. Mary, 25, Woodvale.
Barbour, Florence, 4, Woodvale.
Barley, Mrs. Barbara, 56, Woodvale.
Barley, Nancy, 29, Woodvale.
Barley, Viola, 9, Woodvale.
Beam, Dr. Lemon T., 55, Market street.
Beam, Charles C., 4, Market street.
Beam, Dr. W. C., 35, Locust street.
Beam, Mrs. Clara, 32, Locust street.
Beckley, E. E., 23, Main street.
Bending, Mrs. Elizabeth, 48, Locust street.
Bending, Jessie, 24, Locust street.
Bending, Katie, 15, Locust street.
Beneigh, John C., 65, Cambria.
Benford, Mrs. E. E., 63, Hulbert House.
Benford, Maria, 34, Hulbert House.
Benford, May, 26, Hulbert House.
Benford, Louis, 30, Hulbert House.
Benshoff, J. Q. A., 62, Somerset street.
Benshoff, Arthur, 27, Somerset street.
Bowman, Nellie, 9, Haynes street.
Bowman, Charles H., 7, Haynes street.
Bowman, Frank P., 33, Woodvale.
Bowman, Emma, 28, Woodvale.
Brinkey, Dr. J. C., 28, Franklin street.
Brinkey, Elmer, 26, Hulbert House.
Buchanan, John S., 69, Locust street.
Buchanan, Mrs. Kate J., 63, Locust street.
Buchanan, Robert L., 20, Locust street.
Connelly, Maud, 6, Franklin.
Constable, Philip E., 60, Broad street.
Cope, Mrs. Margaret, 65, Conemaugh.
Cope, Ella B., 28, Conemaugh.
Cooney, Mrs. Elizabeth.
Davis, Mary Ann, 40, Woodvale.
Davis, Thomas S., 59, Locust street.
Davis, Mrs. Elizabeth.
Davis, Mrs. Susan, 27, Millville.
Davis, Clara, 8, Millville.
Davis, Willie, 3, Millville.
Davis, Eliza M.
Davis, Margaret, E.
Davis, Mrs. Cora B., 25, Water street.
Davis, William L.
Davis, Willard G.
Davis, Mary G.
Delaney, Mrs. Jessie, 29, Vine street.
Delaney, Mrs. Ella A.
Dibert, John, 56, Main street.
Dibert, Blanche, 9, Main street.
Dixon, David, 40, Millville.
Diller, Rev. Alonzo P., Locust street.
Diller, Mrs. Marion, Locust street.
Diller, Isaac, Locust street.
Dinant, Lola, Locust street.
Dorris, August.
Drew, Mrs. Mark, 62, Millville.
Drew, Mollie, 8, Conemaugh street.
Duncan, Mrs. Sarah A., 23, Woodvale.
Dyer, Mrs. Nathan, 64, Somerset street.
Eck, Mary Ellen.
Edwards, Mrs. Annie R.
Eldridge, Samuel B., Apple Alley.
Eldridge, Abram S., 34, Merchants' Hotel.
Etchison, John, 44, Napoleon street.
Evans, Mrs. William F., 63, Union street.
Evans, Maggie, 11, Lewis Alley.
Evans, Kate, 5, Lewis Alley.
Evans, Mrs. Josiah, 36, Vine street.
Evans, Maggie, 16, Vine street.
Evans, Lake, 6, Vine street.
Evans, Ira, 6 months, Vine street.

269

Evans, Mrs. Maggie, 37, Vine street.
Evans, Mrs. Ann.
Evans, Sadie, 8, Vine street.
Evans, Herbert, 3, Vine street.
Evans, Pearl, 1, Vine street.
Evans, Lizzie.
Fails, Francis.
Fenn, John, 35, Locust street.
Fenn, Genevieve, 9, Locust street.
Fenn, Bismarck S., 3, Locust street.
Findlay, Lulu, 16. Woodvale.
Fisher, John H., 55, Main street.
Fisher, Mary J., 46, Main street.
Fisher, Emma K., 23, Main street.
Fisher, Ida, 19, Main street.
Fisher, Madge, 10, Main street.
Fisher Minnie, 21, Main street.
Fisher, George, 12, Main street.
Fisher, Frank, 9 months.
Fleck, Leroy Webster.
Fox, Martin, 51, Conemaugh.
Frank, John, Sr., 58, Washington street.
Frank, Mrs. Eliza, 44, Washington street.
Frank, Katie, 19, Washington street.
Frank, Emma, 17, Washington street.
Frank, Laura, 12, Washington street.
Fredericks, Mrs. A. G., 45, Millville.
Fredericks, Mrs. Sarah A.
Frederick, Edmon.
Fritz, Maggie, 26, Conemaugh.
Fritz, Kate, 22, Conemaugh.
Fronheiser, Mrs. Kate, 33, Main street.
Fronheiser, Bessie, 8, Main street.
Fronheiser, Catherine, 3 months, Main street.
Gageby, Mrs. Rebecca, 74, Jackson street.
Gageby, Sadie, 27, Jackson street.
Gallagher, Prof. C. F., 34, Main street.
Gallagher, Lizzie, 29, Main street.
Gard, Andrew, Jr., 25, Main street.
Geddes, George, 47, Woodvale.
Geddes, Marion, 17, Woodvale.
Geddes, Paul, 15, Woodvale.
Gilmore, Mrs. Margaret, 40, Union street.
Gilmore, Anthony, 8, Union street.
Gilmore, Llewelyn, 6, Union street.
Gilmore, Willy, 4, Union street.
Gilmore, Clara, 2, Union street.
Golde, Mrs. Henry, 32, Walnut street.
Griffin, Mary, 47, Walnut street.
Hager, Mary E., 33, Washington street.
Hager, Mrs. Emma.
Hamilton, Jacob, 70, Bedford street.
Hamilton, Jessie, 30, Bedford street.
Hamilton, Laura, 24, Bedford street.
Hamilton, Alex, Jr., 35, Locust street.
Hamilton, Mrs. Alex, 30, Locust street.
Hamilton, Marion, Locust street.
Hamilton, Louther J.
Hammer, George K., 19, Moxham.
Harris, Mrs. William T.
Harris, John, 3, Market street.
Harris, Margaret, 47, Market street.
Harris, Wm. L., 23, Market street.
Harris, Winnie, 21, Market street.
Harris, Maggie A., 19, Market street.
Harris, Sarah, 16, Market street.
Harris, Frank, 12, Market street.
Haynes, Walter B., 22, Horner street.
Haynes, Laura C., 20, Horner street.
Hennekamp, Rebecca, 24, Franklin street.
Hennekamp, Oscar E., 2, Franklin street.
Hennekamp, S. E., 27, Lincoln street.

Heidenthal, Harry R.
Heiser, George, 50, Washington street.
Heiser, Mrs. George, 48, Washington street.
Helsel, George, 16, Johns street.
Hite, Mrs. Ella, 37, Somerset street.
Hochstein, Henry, 30, Conemaugh.
Hoffman, Benjamin F., 56, Market street.
Hoffman, Mrs. Mary, 43, Market street.
Hoffman, Bertha, 19, Market street.
Hoffman, Minnie, 16, Market street.
Hoffman, Marion, 14, Market street.
Hoffman, Florence, 10, Market street.
Hoffman, Joseph, 8, Market street.
Hoffman, Helen, 4, Market street.
Hoffman, Freda, 1, Market street.
Hoffman, Mrs. Mary, 41, Washington street.
Hohnes, Mrs. Ann, 24, Conemaugh.
Hohnes, Mrs. Elizabeth, 80, Lincoln street.
Hohnes, Julia, 18, Conemaugh.
Hollen, Charles.
Howe, Thomas J., 4, Bedford street.
Howells, William, 59, Union street.
Howells, Maggie, 23, Union street.
Howells, Mrs. Ann.
Hughes, Maggie, 22, Sugar Alley.
Hughes, Evan, 57, Sugar Alley.
Humm, Geo. C., Merchants' Hotel.
Humphreys, William, 18, Levergood street.
Jacobs, Lewis, 41, Cambria City.
James, Mrs. Ellen M., 42, Main street.
James, Mollie, 13, Market street.
Jones, Mary J.
Jones, Reuben, 1, Main street.
Jones, James, 32, Conemaugh.
Jones, Ann, 9, Conemaugh.
Jones, Mrs. W. W.
Jones, Edgar R.
Jones, Mrs. Mary A., 52, Pearl street.
Jones, Eliza, 15, Pearl street.
Karns, Joseph, 50, Locust street.
Keedy, Harry C., 30, Millville.
Keedy, Mrs. Mary, 32, Millville.
Kegg, William E., 17, Locust street.
Keiper, Essie J., 24, Franklin.
Keiper, Ralph, 5 months, Franklin.
Kennedy, H. D., 32, Stonycreek street.
Keyser, Mrs. John.
Keyser, Ralph.
Keighly, Mary L., 52, Main street.
Kidd, Joshua, 65, Walnut street.
Kidd, Mrs. Sarah, 60, Walnut street.
Kirkbride, Mahlon, 33, Hager Block.
Kirkbride, Mrs. Ida, 30, Hager Block.
Kirkbride, Luida, 8, Hager Block.
Kirlin, Thomas, 40, Conemaugh street.
Kirlin, Eddie, 12, Conemaugh street.
Kirlin, Frank, 5, Conemaugh street.
Knorr, Mrs. Mary, 45, Jackson street.
Knorr, Emma, 16, Jackson street.
Knorr, Bertha, 14, Jackson street.
Knox, Mrs. Thomas, 45, Somerset street.
Koenstyl, Samuel.
Kratzer, Mrs. Mary, Market street.
Kuntz, Wade, 21, Morris street
Lambreski, Mrs. Barbara, 35, Cambria.
Lambreski, Mary, 6, Cambria.
Lambreski, John, 4, Cambria.
Layton, Mrs. Elvira.
Layton, William, 58, Broad street.
Layton, Mrs. William, 53, Broad street.
Layton, May, 22, Broad street.
Layton, David, Broad street.

Layton, Ella, Broad street.
Lee, Dr. J. K., 48, Main street.
Leitenberger, Mrs. Leah, 68, Vine street.
Leitenberger, Nancy, 48, Vine street.
Leitenberger, Ella, 35, Vine street.
Leitenberger, Eliza, 46, Vine street.
Lenhart, Samuel, 58, Clinton street.
Lenhart, Mrs. Mary, 56, Clinton street.
Lenhart, Annie E., 20, Clinton street.
Lenhart, Emma J., 17, Clinton street.
Lenhart, Katie M., 13, Clinton street.
Lewis, Mrs. Ann.
Lewis, Ananias, 41, Millville.
Levergood, Mrs. Jane, 75, Bedford street.
Levergood, Lucy, 45, Bedford street.
Lewis, Orrie P., 6, Millville.
Lewis, James.
Linton, Minnie, 20, Lincoln street.
Litz, Mrs. John, 74, Morris street.
Llewellyn, Mrs. Margaret, 37, Walnut
   street.
Llewellyn, Annie, 5, Walnut street
Llewellyn, Sadie, 8, Walnut street
Llewellyn, Herbert, 3, Walnut street.
Llewellyn, Pearl, 1, Walnut street.
Luckhart, Louis, 69, Main street.
Luckhart, Mrs. Adolph, 26, Main street.
Ludwig, Charles.
Ludwig, Henry G., 34, Bedford street.
Ludwig, Mrs. Kate, 35, Bedford street.
Mangus, Martha.
Marbourg, Dr. H. W., 56, Market street.
McDowell, Geo., 29, Pearl street.
McDowell, Mrs. Agnes, 33, Pearl street.
McDowell, Lilly, 3, Pearl street.
McDowell, Georgia.
McClelland, Mrs. Jennie, 34, Sherman
   street.
McConaghy, Mrs. Kate, 68.
McConaughy, James P., 72, Walnut street.
McConaughy, Mrs. Caroline M., 65, Wal-
   nut street.
McConaughy, Wallace, 25, Walnut street.
McConaghy, Robert W.
McKee, John, 21, Bedford street.
McKinstry, Mrs. Mary C., 45, Hager Block.
McKinstry, Annie R., 14, Hager Block.
McVay, Lizzie, 20, Locust street
Merle, Elmer E.
Moore, Mrs. Charlotte L.
Meyers, Mrs. Elizabeth, 55, Washington
   street.
Meyers, Mary, 24, Washington street.
Meyers, Mrs. Catherine, 31, Millville.
Meyers, John, 3, Millville.
Miller, Jessie B., 16, Somerset street.
Morgan, Mrs. Charlotte, 49, Millville.
Morgan, Martha, 13, Millville.
Morgan, Minnie, 4, Millville.
Murr, Charles, 41, Washington street.
Murr, Maggie, 14, Washington street.
Musser, Charles, 23, Main street.
Nixon, Mrs. Elizabeth, 39, Woodvale.
Nixon, Emma R., 16, Woodvale.
Nixon, Eddie, 8, Woodvale.
Noro, Kate.
Owens, Gladies, 5 months, Conemaugh
   street.
Owens, Thomas, 10, Conemaugh street.
Owens, William, 65, Market street.
Owens, Annie.
Owens, Mrs. Mary Ann, 31, Conemaugh
   street.

Owens, Mary, 8, Conemaugh street.
Oyler, Mrs. Mary R., 27, Woodvale.
Oyler, John R., 6, Woodvale.
Parke, Mrs. Agnes J., 56, Bedford street.
Parke, William E., Bedford street.
Parsons, Mrs. Eva M., 23, Locust street.
Penrod, William H., 59, Conemaugh.
Peyton, John W., 65, Clinton street.
Peyton, George A., 19, Clinton street.
Peyton, Marcellus K., 16, Clinton street.
Peyton, Julia F., 13, Clinton street.
Phillips, Mrs. Jane M., 68, Market street.
Pike, William W., 50, Haynes street.
Pike, William W., Jr., 15, Haynes street.
Pike, S. Bowen, 10, Haynes street.
Poland, Walter, 5, Market street.
Poland, Frederick, 3, Market street.
Potter, Joseph R., 63, Woodvale.
Potter, Mrs. Sarah, 59, Woodvale.
Potter, Nora G., 17, Woodvale.
Potts, Miss Jane E., 47, Walnut street.
Powell, Richard, 4 weeks, Vine street.
Powell, George, 1½, Vine street.
Pritchard, Henry, 62, Market street.
Prosser, Fannie, 22, Market street.
Prosser, Bessie, 19, Market street.
Prosser, Maria.
Purse, Mary L., Market street.
Raab, George, 44, Clinton street.
Rabb, Mrs. George, 88, Clinton street.
Rabb, Norma, 16, Clinton street.
Raab, Lizzie, 24, Washington street.
Raab, Emilia, 20, Washington street.
Raab, John C.
Raab, Ella.
Rainey, Mrs. Lizzie L., 25, Bedford street.
Rainey, Parke, 1½, Bedford street.
Randolph, George F., 26, Beaver Falls.
Reibert, Julius, Washington street.
Reese, Sarah, 10, Conemaugh street.
Reese, John, 2, Conemaugh street.
Reese, Mrs. J. W.
Reese, Samuel.
Reese, Idris, 3, Vine street.
Reese, Gertie.
Reese, Mrs. Mary D., 74, Market street.
Repp, Mrs. Catherine, 26, Sherman street.
Rhodes, Link, 26, Somerset street.
Rhodes, Ellen, 20, Somerset street.
Rhodes, Clarence, 10 months, Somerset
   street.
Ripple, Jackson, 34, Apple alley.
Roberts, Howard J., 59, Walnut street.
Roberts, Mrs. Howard J., 50, Walnut street.
Roberts, Otis, 23, Walnut street.
Roberts, Mrs. Lucinda H., 81, Main street.
Robinson, Thomas, 60, Woodvale.
Rodgers, Mrs. D. L.
Rodgers, Mrs. Rose, 48, Millville.
Roland, Louis, 31, Conemaugh.
Roland, Lizzie, 29, Conemaugh.
Rosensteel, James M., 50, Woodvale.
Rose, Harry G., 29, Locust street.
Roth, Mrs. Kate, 27, Bedford street.
Roth, John, 38, Potts street.
Schoff, Mrs. E. T., 32, Clinton street.
Schotz, Mrs. Elizabeth, 63, Union street.
Schotz, Annie, 23, Union street.
Schotz, Jennie, 21, Union street.
Schubert, C. T., 39, Stonycreek street.
Seibert, Henry, 58, Woodvale.
Shaffer, Jacob, 47, Cambria.
Shulteis, Henry, 26, Potts street.

Shumaker, John S., 11, Locust street.
Shumaker, Edith M., 7, Locust street.
Shumaker, Irene G., 5, Locust street.
Shumaker, Walter S., 2, Locust street.
Slick, George R., 60, Stonycreek street.
Smith, Mr., 54, Cambria.
Smith, Mrs. Sarah, 72, Walnut street.
Stremel, Julius R., 21, Washington street.
Streum, John, 63, Locust street.
Stufft, J. Wesley, 37, Woodvale.
Stufft, Mrs. J. W., 39, Woodvale.
Suder, Homer, 7. Millville.
St. John, Dr. C. P., 32, Hulbert House.
Stophel, Mrs. Maggie, 21, Baumer street.
Stophel, Frank Earl, 4, Baumer street.
Stophel, Bertha, Hulbert House.
Swank, Mrs Ella, 29, Main street.
Swank, Jennie, 15, Bedford street.
Swank, Jacob, 61, Bedford street.
Swank, Mrs. Catherine, 57, Bedford street.
Swank, Maud, 11, Napoleon street.
Swank, Fred B., 10, Bedford street.
Swank, Susan, 8, Napoleon street.
Swank, Mrs. Neff, 31, Napoleon street.
Swank, Samuel, 5, Napoleon street.
Swank, Edna, 3, Napoleon street.
Statler, Mrs. Amelia, 51, Park Place.
Statler, May, 23, Park Place.
Statler, Frank E., 17, Park Place.
Teeter, Mrs. Mary, 83, Locust street.
Tittle, Cyrus P., 53, Broad street.
Tradenick, Edward, 18, Union street.
Turner, May, 15, Main street.
Tyler, Jno. T., 29, Stonycreek Township.
Thoburn, Thomas, 17, Millville.
Thoburn, Jennie, 7, Millville.
Thomas, Mrs. Mary A., 39, King street.
Thomas, Ida, 7, King street.
Unverzagt, George, Sr., 67, Main street.
Unverzagt, George, Jr., Main street.
Updegraff, Samuel, 15, Woodvale.
Viering, Mrs. Louisa, 38, Conemaugh.
Viering, Lizzie, 20, Conemaugh.
Viering, Henry, 14, Conemaugh.
Viering, Herman, 1, Conemaugh.
Vinton, Margaret, 8, Jeannette, Pa.
Von Alt, Mrs. Catherine, 80, Washington street.
Wagoner, Dr. George, 63, Market street.
Wagoner, Lizzie, 20, Market street.
Wagoner, Mrs. Mary L.
Wagoner, Frances E., 18, Market street.
Wagoner, Cora M., Market street.
Wenner, Carl, 32, Locust street.
Wenner, Mrs., Locust street.
Wenner, Mary, 1, Locust street.
Weaver, Mrs. Sue D., 27, Market street.
Weaver, Martha B., 15 months, Market street.
Weakland, John W., 30, Napoleon street.
Werry, Thomas Albert, 17, Chestnut street.
White, Mrs. Mima.
White, Mrs. Ella, 34, Union street.
White, Mrs. Margaret E.
White, Mary P., 20, Market street.
White, Maggie, 31, Union street.
Wild, Jacob, 72, Main street.
Wild, Mrs. Jacob, 58, Main street.
Wild, Bertha, 16, Main street.
Williams, Maggie, 26, Lewis alley.
Williams, Joseph M., 22, Conemaugh street.
Williams, William J., Union street.
Williams, Carrie E., 20, Woodvale.

Worthington, Mrs. Richard, 28, Conemaugh st.
Worthington, Richard, Jr., 1, Conemaugh st.
Worthington, Mamie, 7, Conemaugh street.
Worthington, Annie, 4, Conemaugh street.
Young, Katie.
Young, Emil, 48, Levergood street.
Young, Frank, 16, Levergood street.
Young, August, 29, Main street.
Young, Andrew C., 36, Broad street.
Zimmerman, Emma, 16, Bedford street.
Zimmerman, Theo. F., 34, Locust street.

SANDYVALE CEMETERY.

Abler, August, 28, Conemaugh.
Abler, Mrs. Louisa, 31, Conemaugh.
Abler, George, 11, Conemaugh.
Baldwin, George, 69, Apple alley.
Bishop, Charles, 45, Woodvale.
Brindle, Mollie, 25, Conemaugh.
Clark, John.
Davis.
Davis.
Davis.
Eberle, Joseph, 63, Conemaugh.
Forbes, Mrs. Rachael, 38, Pearl street.
Forbes, Harry E., 10, Pearl street.
Fredericks, Mrs. Annie E., 78, Vine street.
Gray, S. Taylor, 37, Woodvale.
Gallagher, C. F.
Gallagher, Mrs. C. F.
Greenwood.
Greenwalt, Mrs.
Greenwalt, child.
Given, Jane, Millville.
Given, Benj. F., Millville.
Greenwood, Jennie, 17, Cambria.
Greenwood, Geo., 55, Cambria.
Hammer, Daniel, Railroad street.
Hesselbein, Chas., 27, Conemaugh.
Hesselbein, Lewis, 23, Conemaugh.
Hite, Samuel, 26, Woodvale.
Hoffman, Gottfried, 40, Washington street.
Hoffman, Harry, 6, Washington street.
Hoffman, Daniel.
Hoffman, Godfrey, 41, Washington street.
Hoffman, Lizzie, 16, Washington street.
Hoffman, Mrs. Conrad, 38, Market street.
Hoffman, Charles B., 16, Market street.
Hoffman, Willie, 15, Market street.
Hoffman, Annie, 11, Market street.
Hughes, Emma, 26, Potts street.
Jones, Thomas, 50, Woodvale.
Jones, Mary W., 21, Woodvale.
Jones, Richard, 46, Woodvale.
Jones, child of Richard, Woodvale.
Jones, Clara. 6, Woodvale.
Kimpel, Christ., 47, Clinton street.
Knee, Geo. D., 54, Conemaugh.
Meyers, Mrs. Mary, 69, Cambria.
Morgan, Job, 50, Walnut street.
McClarren, Samuel, 49, Cambria.
McClarren, Mrs. Jane, 42, Cambria.
McClarren, Smith, 22, Cambria.
McClarren, Jno. J., 19, Cambria.
McClarren, James, 4, Cambria.
Peppler, Wm., 20, Conemaugh.
Raab, Geo. C., 28, Washington street.
Reese.
Reese.
Reese.
Reese.

Recke, Alex., 35, Washington street.
Scheetz, Jacob, 61, Clinton street.
Schnable, Conrad, 38, Baumer street.
Schnable, John, 20, Main street.
Stahr, Fred.
Strayer, Mrs. Elizabeth, 47, Market street.
Strayer, Cora, 17, Market street.
Strayer, Bertha, Market street.
Teeter, Mrs.
Thomas, Mrs. Edward, Woodvale.
Thomas, Edward, 49, Woodvale.
Thomas, Lydia, 12, Woodvale.
Thomas, Frank, 8, Woodvale.
Tross, Mrs. Margaret, 39, Woodvale.
Unverzagt, Lizzie, Washington street.
Unverzagt, Minnie, 27, Washington street.
Will, Casper, 45, Bedford street.
Wier, Frank A., 18, Cambria.
Willower, Miss Bella, Somerset street.
Wehn, Mrs. Rachel, 57, Main street.

LOWER YODER CATHOLIC CEMETERY.

Blair, Mrs., 50, Woodvale.
Bopp, Jacob, 32, Broad street.
Bracken, Katie, 21, Woodvale.
Bracken, Minnie, 19, Woodvale.
Bridges, Chas., 2, Cambria.
Bridges, Emma, 18, Cambria.
Brown, Peter, 65, Woodvale.
Brown, Thomas, 24, Woodvale.
Brown, Emma, 20, Woodvale.
Brown, Gertrude, 17, Woodvale.
Byrne, John, 32, Hulbert House.
Byrne, Ella, 24, Hulbert House.
Carroll, Mrs. Bridget, 70, Conemaugh.
Carroll, Thomas, 30, Conemaugh.
Carroll, Rose, 20, Conemaugh.
Clark, Mrs. J. B., 39, Conemaugh.
Cronin, Daniel, 50, Vine street.
Cullen, James, 55, Cambria.
Cullen, Mrs. Ann, 50, Cambria.
Cullen, Mrs. Alice, 48, Cambria.
Cullen, Annie, 20, } Pro'bly dupl'ted
Cullen, Annie, 20, }
{ Cambria.
{ Locust st.
Cush, Mrs. Ann, 55, Cambria.
Cush, Daniel, 33, Cambria.
Cush, Joseph, 19, Cambria.
Cush, Mrs. Tillie, 20, Cambria.
Daily, Mrs. Ann, 60, Locust street.
Daily, Frank, 30, Locust street.
Degnan, Mrs. Mary, 60, Cambria.
Downs, Mrs. Catherine, 55, Millville.
Downs, Mary, 32, Millville.
Downs, Katie, 28, Millville.
Dowling, Mrs. Catherine, 42, Market street.
Dowling, Mary E., 21, Market street.
Dunn, Mary Ann, 25, Prospect.
Early, Mary, 22, Woodvale.
Fitzpatrick, Mrs. Peter, 28, Cambria.
Fitzpatrick, Ella, 6, Cambria.
Fitzpatrick, Mary, 3, Cambria.
Fitzharris, Christ, 42, Franklin street.
Fitzharris, Mrs. Margaret, 40, Franklin street.
Fitzharris, Christ, Jr., 14, Franklin street.
Fitzharris, John, Jr., 12, Franklin street.
Fitzharris, Maggie M., 9, Franklin street.
Fitzharris, Gertie, 5, Franklin street.
Fitzharris, Katie, 7, Franklin street.

Gaffney, Catherine, 2, Cambria.
Gaffney, John, 4, Cambria.
Gallagher, Mrs. Margaret, 32, Washington street.
Gallagher, Thomas, 4, Washington street.
Garvey, Bernard, Sr., 62, Cambria.
Grady, Mrs. Abbie, 60, Cambria.
Halleron, May, 5, Washington street.
Hayes, Mrs. Jane, 32, Cambria.
Hayes, Michael, 12, Cambria.
Hayes, Mary, 8, Cambria.
Hayes, Rose, 7, Cambria.
Hayes, John, 6 Cambria.
Hart, Eliza.
Harrigan, Ella, 22, Hulbert House.
Howard, James B., 45, Conemaugh.
Howe, Mrs. Edward, 50, Railroad street.
Howe, Mrs. Bridget, 48, Cambria.
Howe, Maggie, 24, Cambria.
Howe, Lizzie, 22, Cambria.
Howe, Rose, 19, Cambria.
Howe, Gertrude, 13, Railroad street.
Kane, John, 20, Cambria.
Kane, Mary, 18, Cambria.
Kinney, Mrs. Mary, 50, Washington street.
Kinney, Mary Ellen, 12, Washington street.
Kirby, Wm., 32, Washington street.
Kirby, Mrs. Lena, 25, Washington street.
Lambert, Johanna.
Lavelle, Michael, 22, Broad street.
Lavelle, Wm. M.
Madden, Kate, 17, Cambria.
Matthews, Thos., 22, Clinton street.
McAneny, Neal, 50, Cambria.
McAneny, Mrs. Neal, 45, Cambria.
McAneny, Rose, 23, Cambria.
McAneny, Kate, 18, Cambria.
McAneny, Mary, 13, Cambria.
McAneny, Wm., 9, Cambria.
McAneny, Annie, 5, Cambria.
McAneny, Agnes, 2, Cambria.
McGee, John, 55, Market street.
McGinley, James, 34, Conemaugh.
McVay, Lizzie, 20, Locust street.
Mullin, Peter, 50, Conemaugh.
Murphy, Michael J., 34, Brunswick Hotel.
Murphy, Mrs. Mary, 26, Millville.
Murphy, John, 17, Millville.
Murphy, Rose, 14, Millville.
Murphy, Wm., 11, Millville.
Murphy, J. J., 55, Park Place.
Murphy, Lily, 9, Park Place.
Nightly, John, 30, Millville.
O'Connel, Capt. Patrick, 70, Washington street.
O'Connel, Margaret, 63, Washington street.
O'Connel, Nora, 60, Washington street.
O'Donnel, Frank, Washington street.
O'Neil, Edward, 3 months, Cambria.
O'Neil, Mrs. Bridget, 28, Cambria.
O'Neil, John.
Quinn, Ellen, Franklin street.
Quinn, John, Franklin street.
Riley, Frances, 15, Cambria.
Riley, Gertrude, 13, Cambria.
Riley, Mary, 18, Cambria.
Rogers, Mary, 17, Millville.
Rogers, Tatt.
Rogers, Mrs. Susan.
Rogers, Jane, child.
Ryan, John, 55, Washington street.
Ryan, Mrs. John, 50, Washington street.
Ryan, Maggie, 14, Washington street.
Ryan, Mrs. Mary, 73, Washington street

Sagerson, Catherine, 4, Railroad street.
Sagerson, Agnes, 2, Railroad street.
Sagerson, Thomas, 6 months, Railroad street.
Sharkey, Mary, 4, Washington street.
Sinniger, Mrs. Mary, Cambria.
Slick, Mrs. Nancy, 55, Fourth Ward.
Takacs, Mrs. Teresa, 31, Cambria.
Takacs, Mrs. John, 21, Cambria.
Tokar, Mrs. Dora, 23, Cambria.
Tokar, Mary, 4, Cambria.
Tokar, Annie, 1, Cambria.
Taylor, Frances.

ST. MARY'S CEMETERY.
(Lower Yoder.)

Banyan, Mrs. Rose, 36, Cambria.
Betzler, Mrs. Agnes, 38, Cambria.
Boyle, Charles, Sr., 45, Cambria.
Boyle, Mary, 12, Cambria.
Boyle, Charles, 8, Cambria.
Boyle, Thomas, 7, Cambria.
Brotz, Pancrotz, 55, Cambria.
Brotz, Mrs. Lena, 50, Cambria.
Brady, John, 53, Franklin street.
Brady, Mrs. Julia, 50, Franklin street.
Coby, Elizabeth, Cambria.
Culliton, Mrs. Teresa, 28, Cambria.
Deitrich, Mrs. Amelia, 23, Cambria.
Fish, Lena, 17, Cambria.
Fisher, Ignatius, 59, Cambria.
Fisher, Margaret, 14, Cambria.
Fleckenstein, Mrs. Ann, 25, Cambria.
Fleckenstein, Regina, 2, Cambria.
Gerber, Mrs. Margaret, 41, Cambria.
Gerber, John C., 45, Cambria.
Gerber, Rose, 8, Cambria.
Gerber, Vincent, 6, Cambria.
Hanki, Edward, Cambria.
Hecker, Mrs. Christ, 58, Cambria.
Heider, Mrs. Ella, 24, Cambria.
Heider, John Leo, 6 months, Cambria.
Hessler, Mrs. Fedora, 29, Cambria.
Hessler, Mary, 10, Morrellville.
Hessler, Joseph, 1½, Cambria.
Hirsch, Eddie, 8, Cambria.
Just, Magdalena, 29, Cambria.
Just, William, 9, Cambria.
Just, Eddie, 4, Cambria.
Kintz, Mrs. Mary, 26, Cambria.
Kintz, Katie, 19, Cambria.
Kintz, Mrs. Mary, 25, Cambria.
Knoblespeice, Maggie.
Koebler, Mrs. George, 60, Cambria.
Kropp, Katie, 21, Cambria.
Lambert, Johanna, 19, Washington street.
Lambreski, Kate, 12, Cambria.
Macheletzky, Stanislaus, 10, Cambria.
Martinades, Mrs. Mary, Cambria.
Miller, Mrs. Annie M., 46, Cambria.
Miller, George, 65, Cambria.
Miller, Eddie, 3, Cambria.
Miller, Annie, 1, Cambria.
Nich, Mrs. Margaret, 30, Cambria.
Nich, Frank, 6, Cambria.
Nich, John, 4, Cambria.
Osterman, Joseph, 38, Cambria.
Quinn, Mrs. Terry, 26, Railroad street.
Schnell, Mrs. Fidel, 68, Cambria.
Schnell, Mrs. Margaret, 60, Cambria.
Schnell, Mrs. F., Cambria.
Schmitt, Mary, 31, Cambria.
Schmitt, George, 4, Cambria.

Schmitt, Sophia, 1½, Cambria.
Shmitt, Fredericka, Cambria.
Shmitt, Mrs. Hortena, Cambria.
Shmitt, Leo, Cambria.
Sininger, Mrs. Mary, Cambria.
Sarlouis, Mrs. Barbara, 48, Cambria.
Sarlouis, Mrs. Peter, Cambria.
Snell, Mary, 13, Cambria.
Stinely, Mrs. Mary, 35, Cambria.
Stinely, Kate, 12, Cambria.
Stinely, Joseph, 5, Cambria.
Weber, Mrs. Tresa, 43, Cambria.
Weber, John, 4, Cambria.
Weinzierl, Louis, 41, Cambria.

OLD CATHOLIC GRAVEYARD.
(Conemaugh Borough.)

Akers, Alvar, 54, Upper Yoder.
Coad, Mrs. Mary, 57, Washington street.
Coad, John, 59, Washington street.
Conrad, William, 26, Woodvale
Halleran, Mrs. Mary C., 30, Washington street.
Hannan, Eugene, 14, Woodvale.
Howe, Abner.
Lynch, John, 27, Conemaugh.
Lynch, Mary, 16, Conemaugh.
Mayhew, Jennie, 18, Woodvale.
Mayhew, Joseph, 16, Woodvale.
Mayhew, Annie, 12, Woodvale.
Mayhew, Earnest, 9, Woodvale.
Mayhew, Harry, 6, Woodvale.
Mayhew, James, 3, Woodvale.
McKarley, Mrs. Mary.
Nugent, Mrs. Mary Jane, 50, Hager Block.
Quinn, Vincent, 14, Main street.
Wehn, Mrs. Laura, 29, Conemaugh.
Wehn, Annie, 4, Main street.
Wehn, Mary, infant, Conemaugh.
Wehn, Joseph, 4, Conemaugh.
Wheat, Frank, 28, Clinton street.

GERMAN CATHOLIC CEMETERY.
(Sandyvale.)

Brindle, Mary.
Geis, Mrs. Abbey, 24, Salina, Kansas.
Geis, Richard P., 2, Salina, Kansas.
Hable, John, 29, Conemaugh.
Hoffgard, Conrad, 18, Clinton street.
Holtzman, Joseph, 35, Woodvale.
Horne, William J., 21, Conemaugh.
Horne, Emma J., 22, Stormer street.
Hornick, John P., 26, Conemaugh.
Hornick, Mrs. Amelia, 25, Conemaugh.
Horton, Joseph, Sr., 59, Woodvale.
Keifline, Mrs. Catherine, 56, Conemaugh.
Maloy, Manassas, 45, Clinton street.
Malzi, Jacob, 34, Washington street
Murtha, James, 65, Conemaugh.
Murtha, James, 28, Main street.
Murtha, Mrs. Barbara, 24, Main street.
Murtha, Frank, 6, Main street.
Murtha, Flora May, 4, Main street.
Murtha, Lily, 1, Main street.
Oswald, Charles, 44, Third Ward.
Oswald, Mary, 19, Third Ward.
Quinn, Vincent D., 16, Main street.
Ripple, Maggie B., 27, Merchants' Hotel.
Robine, Christina, 25, Franklin street.
Sarlouis, Sophia.
Schnurr, Charles, 40, Conemaugh.

Schnurr, Robert, 27, Smith alley.
Schry, Joseph, Sr., 78, Woodvale.
Schry, Mrs. Joseph, 58, Woodvale.
Shellhammer, Lorentz.
Shellhammer, Patricius.
Schaller, Joseph, 62, Woodvale.
Schaller, Mrs. Joseph, 62, Woodvale.
Schaller, Annie, 24, Woodvale.
Schaller, Rose, 21, Woodvale.
Werberger, Prof. F. P., 70, Locust street.
Voegtly, Germanus, 62, Conemaugh.

(Geistown.)

Rubritz, Peter, 65, Franklin Borough.
Rubritz, Mrs. Margaret, 56, Franklin Borough.
Rubritz, Maggie, 20, Franklin Borough.
Schiffhauer, John, 62, Washington street.
Stenger, John, 12, Main street.
Stenger, Leo, 3, Main street.
Steigerwald, William, Conemaugh.
Steigerwald, Mrs. Mary, 38, Conemaugh.
Steigerwald, infant, 1 month, Conemaugh.

PUBLIC PLOT.

[*Known to have been found, but bodies never recovered by friends, and buried in Public Plot in Grand View Cemetery.*]

Arthur, Earl H., 8, Water street.
Baker, son of Andrew.
Bohnke, Charles.
Bopp, son of Jacob.
Bopp, Katie, 9, Broad street.
Bricker, Henry.
Burns, John.
Barbour, Harry L., 16, Locust street.
Barker, Mrs. Susan, 28, Woodvale.
Behnke, Charles.
Bloch, Louisa, 17, Conemaugh.
Boehler, Mrs. Annie, 39, Conemaugh.
Brawley, George D., 17, Cor. Union and Vine sts.
Brennan, Mrs. Martha, 36, Woodvale.
Brennan, Mary, 16, Woodvale.
Brennan, William, 12, Woodvale.
Brennan, Lewis, 10, Woodvale.
Brennan, Arthur, 7, Woodvale.
Brennan, Frank, 3, Woodvale.
Brown, Sadie, 22, Woodvale.
Bruhn, Claus, 58, Conemaugh.
Bryan, Wm. A., 45, Mansion House.
Campbell, Peter, 40, Conemaugh.
Casey, William, 48, Cambria.
Cornelison, Maggie.
Craig, Thomas A., 32, Market street.
Craig, Mrs. T. A., 30, Market street.
Craig, Christ, 45, Cambria.
Craig, Annie, 13, Walnut street.
Cunz, Lydia, 6, Napoleon street.
Cunz, Robert, 4 months, Napoleon street.
Dillon, James, 35, Napoleon street.
Downey, Mrs. Mary, 55, Pearl street.
Dudzik, Andrew, 28, Cambria.
Eager, Annie.
Eck, Mary Ann, 37, Woodvale.
Eck, Lily, 12, Conemaugh street.
Edwards, Mrs. Ann R., 70, Union street.
Elsaesser, Andrew, 16, Conemaugh.
English, Joseph, 24, Railroad street.
Fagan, Matthew, 40, Millville.
Fagan, Mrs. M., 38, Millville.

Fagan, Monica, 12, Millville.
Fagan, Daniel, 10, Millville.
Fagan, Clara, 3, Millville.
Fagan, Thomas, 1, Millville.
Faloon, Mrs. Ann E., 63, Pearl street.
Fichtner, Mrs. Tillie, 33, Main street.
Fiddler, Elmira, Bedford street.
Fiddler, Eliza J.
Fockler, Herman, 21, Franklin street.
Griffin, Miss Mary.
Hamilton, Mary, 33, Bedford street.
Hanki, Mrs. Teresa, 40, Cambria.
Hause, Mollie.
Hellriggle, Chas.
Hellriggle, Mrs. Lizzie, 30, Woodvale.
Henry, William, 34, Cumberland, Md.
Hocker, Mrs. John, 72, Somerset street.
Hop Sing, Franklin street.
Hurt, Charles, London, England.
Irwin, Maggie, 22, Hulbert House.
Johnson, David, 45, Conemaugh street.
Jones, Mary, 14, Main street.
Kast, Clara, 17, Clinton street.
Keene, Katie, 16, Union street.
Keinxstoel, Samuel, 30, Market street.
Larimer, James, 45, Somerset street.
Lee Sing, Chinaman, Franklin street.
Lucas, Maria, 50, Conemaugh.
Madden, Mrs. Mary, 47, Cambria City.
Mack, August.
McClarren, Cora, 8, Cambria City.
McCue, Mrs.
Melden, Richard.
Maley, Henry.
Mosser, Mrs. Mary, 65, Conemaugh street.
Mullen, Margaret.
Oswald, Mrs.
Owens, Mrs.
Oyler, John R.
Phillips, Mrs. Eliza, 48, Union street.
Reese, Mrs. Lizzie, 30, Conemaugh street.
Reese, Annie, 7, Vine street.
Reidel, John C., 60, Conemaugh.
Rich, Mrs. Charlotte, 45, Stonycreek street.
Roberts, Mrs. Jennie, 18, Somerset street.
Rosenfelt, Solomon, Washington street.
Saylor, Henry.
Schnable, Mrs. Conrad, 35, Baumer street.
Schittenhelm, Anton, Cambria.
Schittenhelm, Anton Jr., Cambria.
Shumaker, Mrs. James M., Locust street.
Skiba, Mrs. Stainslous, 32, Cambria.
Skiba, Joseph, 4, Cambria.
Smith, Ralph, 11, Woodvale.
Smouse, Jennie, Hulbert House.
Stern, Bella.
Strauss, Moses, 77, Vine street.
Strauch, Henry, 50, Conemaugh.
Smith, Willie, 1, Cambria.
Surany, David.
Thomas, John T.
Till, Arthur, 27, Market street.
Unverzagt, Daniel, 66, Washington street.
Unverzagt, Mrs. Daniel, 62, Washington street.
Viering, Mrs.
White, Mrs. John, 76, Union street.
Wagnor, Henry, Cambria.
Warsing, Jane, 24, Coopersdale.
Warkeston, Miss.
Weinzierl, Mrs. Mary, 38, Cambria.
Wearn, Willie, 6, King street.
Walford, Frank.
Will, Elizabeth, Conemaugh.

PASSENGERS ON DAY EXPRESS.

*[Those marked ? bodies never found.
Those found lived at the places named, to
which places the remains were taken.]*

Bates, Mrs. Annie, Delavin, Wis.
? Brady, Mrs. J. W., Chicago, Ill.
Bryan, Elizabeth M., 20, Philadelphia.
Christman, Mrs. A. C., Dallas, Texas.
Day, John R., 60, Prospect, Md.
Day, Miss, Prospect, Md.
Ewing, Andrew, Snow Shoe, Pa.
? Feustermaker, Victor, Egypt, Lehigh
County
Harnish, Blanche, Dayton, O.
? Hemingway, Fred, and wife, Kokomo,
Ind.
King, Mrs. J. F.
? Lyon, E., New York.
? McCoy, Mrs.
? McCoy, ——
? McCoy, ——
Meisel, Christ, 32, Newark, N. J.
Minich, Kate, Fostoria, Ohio.
Paulson, Jennie, 20, Allegheny City.
? Phillips, Frank (porter), Jersey City.
Rainey, Mrs. Sophia, 64, Kalamazoo, Mich.
Ross, John D.
Schrantz, George, Pleasant Gap, Pa.
? Shelly, W., Newark, N. J.
Shick, Cyrus, Reading.
? Sible, Mrs. Springtown, Bucks County,
Pa.
Smith, Mrs. H. K., 25, Osborn, Ohio.
Smith, R. Wardwell, 3, Osborn, Ohio.
Stinson, Eliza, Norristown, Pa.
? Swaney, Mrs. Mary A., 67.
Swineford, Mary A., St. Louis, Mo.
Swineford, Mrs. Ed., St. Louis, Mo.
Tarbell, Mrs. Farney, 32, Cleveland, Ohio.
? Tarbell, Grace, 7, Cleveland, Ohio.
? Tarbell, Bertie, 5, Cleveland, Ohio.
? Tarbell, Howard, 2, Cleveland, Ohio.
Weaver, Beneval, Millersburg, Pa.
Woolf, Jennie, Chambersburg, Pa.

MISCELLANEOUS.

*[Bodies taken to places named in sub-
heads for burial. The place named in line
with name of individual is where they were
lost from.]*

LOYSBURG, BEDFORD COUNTY, PA.
Aaron, Mrs. H. B., 29, Railroad street.
Aaron, Flora, 10, Railroad street.
BLAIRSVILLE, PA.
Alexander, John G., 45, Woodvale.
Alexander, Mrs. John G., 45, Woodvale.
Brown, Emma, 20.
McLaughlin, Mrs. Julia, 60, Cambria.
Miller, Robert, 22, Sixth Ward.
Pike, Fanny, 19, Haynes street.
WILMORE, PA.
Beiter, Mathias, 3, Clinton street.
PHILADELPHIA, PA.
Butler, Chas. T., Hulbert House.
Carlin, Jonathan, Hulbert House.
Cox, James G., Hulbert House.
Clark, W. H. L., 50, Hulbert House.
De Walt, Chas. B., 36, Hulbert House.
Dorsey, John D., Hulbert House.
Lichtenberg, Rev. John, Locust street.
Lichtenberg, Mrs.
Murray, James, 50, Hulbert House.

Nathan, Adolph, 40, Main street.
Overbeck, William H., 38, Main street.
Spitz, Walter L., Hulbert House.
Woolf, Mrs. M. L., Jackson street.
LOUISVILLE, KY.
Marshall, Chas. A., 34, Hulbert House.
BRADDOCK, PA.
Cadogan, Mrs. Mary A., 46, Millville.
Cadogan, Ann, 25, Millville.
Young, Mrs. Kate, 34, Market street.
Young, Samuel, 13, Market street.
PITTSBURGH, PA.
Creed, David, 60, Washington street.
Creed, Mrs. Eliza, 55, Washington street.
Creed, Maggie, 28, Washington street.
Fisher, Moses, 24, Mansion House.
Sweeney, Mrs. Ann, 70, Conemaugh.
BENSHOFF'S, CAMBRIA COUNTY, PA.
Custer, William H., 35, Millville.
STEUBENVILLE, OHIO.
Davis, Frank B., 40, Main street.
Davis, Frank, infant.
SOMERSET, PA.
Gaither, Harry, 18, South street.
Houston, Minnie, Hulbert House.
Hurst, Nathaniel, 15, Washington street.
SHIPPENSBURG, PA.
Diehl, Carrie, 20, Hulbert House.
Wells, Jennie, 22, Hulbert House.
MERCER COUNTY, PA.
De France, Mrs. H. T., 32, Hulbert House.
NEW YORK, N. Y.
Dow, W. F., Hulbert House.
HOLLIDAYSBURG, PA.
Fitzharris, John, Sr., 97, Franklin street.
PHILIPSBURG, PA.
Eskdale, James, 42, Woodvale.
Eskdale, Mrs. James, Woodvale.
BERLIN, PA.
Garman, Grace, 21, Washington street.
PITTSTON, PA.
Groff, Nellie C., 20, Hulbert House.
LEECHBURG, PA.
Hill, Ivy, 6, Washington street.
Jack, Jennie.
COVER'S HILL, CAMBRIA COUNTY, PA.
Hinchman, Harry, 4, Woodvale.
Long, Samuel.
Shaffer, Fred, 21, Conemaugh.
DERRY, PA.
Jackson, H. A., 36.
CUMBERLAND, MD.
Katzenstein, Mrs. Ella, Hulbert House.
Katzenstein, Edwin, Hulbert House.
HARRISBURG, PA.
Keis, Charles A., 26, Conemaugh.
Weber, E. Vincent, 26, Woodvale.
Weber, Mrs. Florence, 25, Woodvale.
BUTLER, PA.
Bonner, Mrs. Ann, 24.
Kenna, Mrs. Alice B.
HUNTINGDON COUNTY, PA.
McDivitt, Mattie, 32, Water street.
HEADRICK'S, CAMBRIA COUNTY, PA.
Allison, Florence, 12, Texas.
Beck, William J., 30, Woodvale.
Beck, Mrs. Blanche, 29, Woodvale.
Wilson, Dr. J. C., 53, Franklin.
Wilson, Caroline E., 52, Franklin.
QUAKERTOWN, PA.
Smith, Mrs. J. L., 34, Hulbert House.
Smith, Florence, 9, Hulbert House.
Smith, Frank, 7, Hulbert House.
Smith, infant, 4 months, Hulbert House.

Wilson, Charles H., 45, Hulbert House.
ARMAGH, PA.
Young, Sarah C., 66, Court street.
BEAVER FALLS, PA.
Leslie, John S., 30, Levergood street.
YPSILANTI, MICH.
Richards, Carrie, Hulbert House.
Richards, Mollie, Hulbert House.
BALTIMORE, MD.
Goldenberg, Henry, 54, Lincoln street.
Hoopes, Walter E., 30, Woodvale.
Smith, Mrs. Alice M., 29, Woodvale.
GREENSBURG, PA.
Kilgore, W. Alex., 52, Washington street.
Montgomery, Alex., 55, Stonycreek street.
SEWICKLEY, PA.
Little, John A., 43, Hulbert House.
BANGOR, PA.
Llewellyn, Mrs. J. J., 27, visiting at J. T.
Llewellyn's.
SOUTH FORK, PA.
Mullin, James, 24.
JERSEY HEIGHTS, N. J.
Myer, Bernhart.
ROME, N. Y.
Richards, John O., 70.
YOUNGSTOWN, OHIO.
White, Mrs. Alex., 42.
INDIANA, PA.
Ziegler, James B., 24.
READING, PA.
Fediman, W. M., 56, Main street.
BLOUGH'S, STONYCREEK TOWNSHIP, PA.
Blough, Samuel, 40, Market street.
Blough, Sophia, 38, Main street.
Blough, child, Main street.
SCALP LEVEL, PA.
Owens, William L., 11, Market street.
Owens, Daisy, 13, Market street.
NICHOLSON, PA.
Rosensteel, Mrs. J. M., 35, Woodvale.
Rosensteel, Ray Halstead, 18, Woodvale.

NO CEMETERY RECORD.

[*Bodies found, but not known where buried.*]

Adams, Henry Clay.
Alberter, Anna, 22, Cambria.
Amps, Nicodemus, 42, Cambria.
Amps, Mrs. Teresa, 32, Cambria.
Atkinson, John, 72, East Conemaugh.
Baer, Rosa L., 17, Grubbtown.
Bagley, William, Morrellville.
Baird, Charles.
Baker, Mrs. Nelson.
Baker, Mrs. Mary, Woodvale.
Baker, Catherine, 70, Market street.
Baker, Agnes, 68, Market street.
Barley, Myrtle, 11, Woodvale.
Barley, Mamie, 7, Woodvale.
Barley, Effie, 5, Woodvale.
Barley, Laura, 6 months, Woodvale.
Barrett, Jas., 27, Franklin st., St. Charles
Hotel.
Berg, Mrs. Marion, 24, Woodvale.
Berkebile, Mahlon, Morrellville.
Blough, Emanuel, 22, Bedford street.
Blough, infant, School alley.
Bowersox, Frank, 22, Market street.
Boyer, Solomon, 62, East Conemaugh.
Bradley, Thomas, 42, Conemaugh.
Bruhn, Mrs. Anna, 45, Portage street.
Bunting, Mrs. Caroline, 45, Woodvale.

Burk, Mrs. Matilda, 38, East Conemaugh.
Burkhard, Mrs. Mollie, 36, Woodvale.
Carr, Alexander, 36, East Conemaugh.
Carr, Sissie, 2, East Conemaugh.
Christie, Andrew C., 50, Woodvale.
Clark, Thomas, 42, Union street.
Clark, John B., 50, Portage st., and 7 children.
Cole, John, Cambria.
Connors, Mrs. Mary, Millville.
Cooper, Otto, 8, Kurtz alley.
Cooper, Mrs. 38, Kurtz alley.
Couthamer, Mr.
Coy, Mrs. Sarah, 46, East Conemaugh.
Coy, Newton G., 16, East Conemaugh.
Craig, Mrs. Catherine, 40, Walnut street.
Crowthers, infant, 3, Chestnut street.
Cummings, Amy, Somerset street.
Davis, Frank, 8, Woodvale.
Davis, Mrs. Philip, 60.
Davis, Mrs. Thomas S., 55, Market street.
Delaney, Mrs. C. W., 59, Conemaugh street.
Dimond, Frank, 36, Conemaugh.
Dimond, Mrs. Ann, 64, Conemaugh.
Doorocsik, Mrs. Annie, 28, Cambria.
Doorocsik, Miss, 6, Cambria.
Doorocsik, Mary, 4, Cambria.
Dorriss, August, 54, Conemaugh.
Doubt, Mrs. William, 63, Cambria.
Dougherty, Mary, 16, Cambria.
Eberle, Lena, 14, Woodvale.
Fails, Dolly F., 15, Union street.
Fers, Frank, 23, Millville.
Fink, Mary E., 17, Conemaugh street.
Fisher, Wolfgang, 33, Main street.
Fisher, Noah, East Conemaugh.
Flegle, David G.
Flegle, Miss Annie.
Flinn, Mrs. Mary, Bedford street.
Fogarty, Thomas, 50.
Forrest, Frank, 12, Locust street.
Foust, Conrad, Woodvale.
Gardner, Rose, 20, Prospect.
Gill, William, 7, Prospect.
Gillen, Laura, Bedford street.
Gordon, Susan L., 62, Hager Block.
Greenwood, Mrs. Rose, 33, Conemaugh.
Gromley, Lilly, 19, Mineral Point, Pa.
Gromley, J. A., 14, Mineral Point, Pa.
Hallen, Charles E., 33, Millville.
Harris, Mrs. Mary T., 48, Walnut street.
Hartzell, Mr., Market street.
Hecker, John, 10, Cambria.
Heckman, Francis, 25, Main street.
Heffley, Edward, 22, Somerset street.
Heine, Henry, 26, Cambria.
Heine, Mrs. Lizzie, 25, Somerset street.
Herman, Edward, Cambria.
Hess, William B., 55, Millville.
Hipp, Elizabeth P., 20, Main street.
Hitchins, Mrs. Cordelia, 35, Market street.
Hornick, Agnes, Broad street.
Hughes, Mary, 7, Chestnut street.
Hughes, Mrs., 64, Union street.
James, Lena, 26, River avenue.
James, Maggie, 1, River avenue.
Jenkins, Mrs. Susan, 40, Somerset street.
Johill, Joseph, Third Ward.
Johnson, John M., 40, Union street.
Johnson, Mrs. John M., 38, Union street.
Johnson, Mrs. Oliver, 22, Conemaugh street.
Jones, Maggie, 29.
Kane, John, 45, Union street.
Kane, Bridget, 20, Market street.

Keifline, Mary, 4, Conemaugh.
Kelly, Charles, Millville.
Kunkle, Lizzie, 21, Washington street.
Laban, Mrs. Teresa, 50, Cambria.
Leech, Mrs. Sarah E., 60, Franklin.
Leech, Alice M., 18, Franklin.
Lingle, Mrs. Mary J., 44, Pearl street.
Long, Samuel, 60, Vine street.
Lotz, Conrad, 64, Sherman street.
Lyden, Mary, 20, Merchants' Hotel.
Maneval, Clarence, 17, Lincoln street.
Mann, Michael, 41, South Fork.
Marczi, Mrs. Mary, 42, Cambria.
Marshall, Wm. H., 23, Clinton street.
Maurer, John, 77, Morris street.
McAuliff, Laura, 16, Woodvale.
McDowell, Geo., 8, Pearl street.
McGuire, Mrs. Mary, 45, Walnut street.
McHugh, Mrs. D. A., 45, E. Conemaugh.
McHugh, Gertrude, 16, E. Conemaugh.
McHugh, Jno. L., 14, E. Conemaugh.
McNally, Patrick, 42, Prospect.
Mecke, August, 51, Cambria.
Melczer, Frederick, 28, Cambria.
Miller, Robert, 5, Napoleon street.
Miller, John A., 25, Cambria.
Miller, William, 44, Franklin.
Mingle, Sarah.
Monteverde, Mary, 11, Washington street.
Monteverde, Emelia, 7, Washington street.
Morran, James A., 53, Somerset street.
Nau, Katie, 20, Bedford street.
Neary, Mrs. Kate, 34, Bedford street.
Neary, Mary Ellen, 11, Bedford street.
Noblespiece, Maggie, 14, Morrellville
Nugent, Mrs. Mary Jane, 60, Hager Block.
O'Connell, —, Cambria.
O'Conner, Rose, 20, Locust street.
O'Neal, John, 19, Wood alley.
Oswald, Appahmarian, 12, Cambria.
Page, Emma, 11, Mineral Point.
Page, Herman, 6, Mineral Point.
Palmer, Mrs. J. H., 76, Napoleon street.
Partsch, Mrs. Josephine, 59, Woodvale.
Phillips, John, 15, Union street.
Rausch, John, 44, Daisytown.
Repp, George, 5 months, Daisytown.
Robine, Eddie, 2, Franklin.
Robine, Willie, 9 months, Franklin.
Rodgers, Patrick, 52, Millville.
Rodgers, Grace, 5, Millville.
Rodgers, Mrs. Mary, 50, Millville.
Ross, Joseph, 30, Conemaugh.
Roth, Annie, 5, Cambria.
Rowland, Emma, 32, Market street.
Rowland, Ran, 16, Market street.
Samen, Mrs. Annie, 25, Cambria.
Samen, John, 4 Cambria.
Samen, Annie, 3, Cambria.
Samen, Mary, 3 months, Cambria.
Schmidt, Mrs. Frederick, Cambria.
Schmidt, Hortense.
Schmidt, Leo.
Schmitz, Gustave, 33, Clinton street.
Schittenhelm, Max, Cambria.
Snyder, Mrs. Annie, 34, Woodvale.
Spareline, John, 64, Railroad street.
Smith, Mrs. Maggie L., 38, Woodvale.
Smith, Addie, 13, Pearl street.
Smith, Philip, 16, Walnut street.
Smith, Mrs. Amelia, 32, Cambria.
Smith, Mrs. Mary, 52, Conemaugh.
Smith, Philip, 16, Conemaugh.
Slick, Josephine, 20, Woodvale.

Sutliff, George, 25, Somerset street.
Stern, Bella, 1, Washington street.
Stewart, —, Second Ward.
Spicsak, Mrs. Annie, 27, Cambria.
Tacey, Peter L., 20, Woodvale.
Trindle, John M., 39, Nineveh.
Trawatha, Mrs. Annie, 60, Conemaugh.
Thomas, Mabel, 6, Market street.
Thomas, Edward M., 71, Woodvale.
Uhl, Mrs. Ludwig, 80, Peter street.
Valentine, George M., 42, Market street.
Weisz, Mrs. Martin, 46, Cambria.
Weisz, Jacob, 13, Cambria.
Weisz, Jacob.
Weisz, Isaac, 6, Cambria.
Weisz, Anna, 4, Cambria.
Welsh, Thomas, 60, Cambria.
Weinzierl, Louis, 41, Cambria.
Williams, Elanor, 7 months, Lewis alley.
Williams, Mrs. Margaret, 27, Conemaugh
street.
Wild, Mrs. Margaret, 80, Conemaugh.
Willower, Miss Bella, Somerset street.
Willower, Bertha, Somerset street.
Wissinger, Mrs. Catherine, 47, Morris street.
Yost, Charlotte, 16, Pine street.

NOT KNOWN TO HAVE BEEN FOUND.

Abele, Katie, 21, Main street.
Abler, Lulu, Woodvale.
Alberter, Teresa, 3, Cambria.
Alexander, Mrs. Martha, Main street.
Allison, Mrs. Jane, 45, Pittsburgh.
Alt, John, 65, Conemaugh.
Alt, Teresa, 20, Conemaugh.
Alt, George, 60, Cambria.
Alt, Mrs. Ann, 75, Cambria.
Amps, Mary, 11, Cambria.
Aubrey, Thomas, 45, Conemaugh street.
Backer, George, 27, Conemaugh.
Baker, James, 22, Woodvale.
Baker, Catherine, Market street.
Baker, Lydia, 20, Woodvale.
Baker, Nancy, Market street.
Baker, Richard, 1, Woodvale.
Baker, Mellville, 11, Woodvale.
Baker, Deronda, 5, Woodvale.
Baker, Dolly, Woodvale.
Baker, Clara, 17, Woodvale.
Banyan, John, 7, Cambria.
Banyan, Albert, 4, Cambria.
Banyan, Theodore, 2, Cambria.
Barbour, Howard, 7, Woodvale.
Barbour, John F., 3 months, Woodvale.
Barbour, Mrs. Sarah, 59, Woodvale.
Barker, Edward, 27, Woodvale.
Barker, Clara, 2½, Woodvale.
Barker, infant, 1 month, Woodvale.
Bartosh, Mrs. Hannah, 39, Cambria.
Bartosh, Frank, 14, Cambria.
Baumer, Mrs. Eliza, 68, Woodvale.
Beam, Roscoe, 2, Locust street.
Beecher, Mrs. Jane, 44, Woodvale.
Beecher, Mary, 23, Woodvale.
Beck, Alfred, 6.
Beck, Roy, 3.
Beckley, Mrs. Mary, 48, Woodvale.
Benson, Mrs. Bessie, 23, Cambria.
Benson, Flora, 3, Cambria.
Bare, Mrs.
Bare, infant.
Barkley, George.
Barnes, Andrew, Conemaugh.

Barron, Anton.
Barron, Mrs.
Benson, Cora Belle, 1½, Cambria.
Berkey, Henry S., 45, Clinton street.
Beske, John, 7, Cambria.
Beske, Joseph, 5, Cambria.
Beske, Frank, 3, Cambria.
Beske, Lewis, 1, Cambria.
Betzler, Frank, 9, Cambria.
Betzler, Katie, 7, Cambria.
Bishop, Julius, 55, Cambria.
Bitner, A. B.
Blair, Alfred, 53, Woodvale.
Blair, Oliver, 25, Woodvale.
Blair, Alfred, Jr., 14, Woodvale.
Blair, Emanuel, 12, Woodvale.
Blair, Rosana, 10, Woodvale.
Bloch, Mrs. Rose, 54, Conemaugh.
Bloch, Annie, 26, Conemaugh.
Block, Minnie, 15, Conemaugh.
Bloch, Emma, 13, Conemaugh.
Boehler, Barbara, 7, Conemaugh.
Boehler, Annie, 9, Conemaugh.
Bogus, William.
Blough, Mrs. First Ward.
Bopp, Naomie, 7, Broad street.
Bonson, Charles R.
Booser, Eddie, 14, Market street.
Bowers, George, Woodvale.
Bowersox, Mrs. Elia, 16, Market street.
Bowersox, Cordelia, 3, Market street.
Bowman, Jessie, 4, Woodvale.
Bowman, Blanche, 2, Woodvale.
Boyer, Emma, 17, Woodvale.
Boyle, Rose, 6, Cambria.
Boyle, Bridget, 4, Cambria.
Boyle, William, 2, Cambria.
Boyle, Joseph, 8 months, Cambria.
Braden, Patrick, Millville.
Bradley, Mrs. Elvira, 39, Conemaugh.
Brawley, Mrs. Maggie, 42, Union street.
Brawley, Robert J., 4, Union street.
Brawley, John.
Brennan, Mrs. Mary Ann, 46, Woodvale.
Brennan, Mary Ann, 23, Woodvale.
Brennan, Ellen, 19, Woodvale.
Brennan, Jane, 16, Woodvale.
Brennan, Agnes, 13, Woodvale.
Bridges, Mrs. Jane, 64, Market street.
Brindle, Vincent.
Brindle, Frank.
Brindle, Rose.
Brinker, Henry.
Briney, Matilda, 25. Woodvale.
Brockner, Samuel, 28, Conemaugh.
Brown, Mrs. Magdalena, 58, Cambria.
Brown, Lizzie, 15, Woodvale.
Brown, Mrs., Conemaugh.
Buckhard, Mrs. Elizabeth, 50, Woodvale.
Buckhard, Charles, 19, Woodvale.
Buckhard, Mrs., 63, Woodvale.
Buckley, Mrs. Mary, 48, Woodvale.
Burket, Frank, 14, Washington street.
Burket, Blair, 8.
Burkhard, Howard, 12, Woodvale.
Burkhard, Gussie J., 5, Woodvale.
Burkhard, Charles C., 2, Woodvale.
Burkhard, Mrs. Catherine, 85, Mineral Point.
Burns, Peter, Woodvale.
Butler, John, 51, 84 John street.
Butler, Robert, 40, Millville.
Butler, Mrs., 70, Millville.
Butler, Annie, 17, Millville.

Butler, Fannie, 14, Millville.
Butler, George, 11, Millville.
Butler, Mrs. Sarah.
Byers, Mrs. Catherine, 46, Mineral Point.
Callahan, Mary, 22, Locust street.
Callahan, Mrs. Frank, Locust street.
Carr, Mrs. Mary, 42, Woodvale.
Carr, William, 7, Woodvale.
Carr, Patrick, 22, Cambria.
Carr, Mrs. Sarah, 20, Cambria.
Cartin, Mrs. Thomas, 46, Woodvale.
Cartin, Frank, 5, Woodvale.
Christie, Mrs. Lizzie, 46, Woodvale.
Christie, Daisy, 19, Woodvale.
Clark, Thomas, Jr., 9, Union street.
Clark, Annie, 5, Union street.
Clark, Hamilton.
Coad, William, 12, cor. Market and Washington.
Cleary, Alice, Cambria.
Conrad, John, 21, Woodvale.
Constable, Mrs. Sarah E., 48, Broad street.
Constable, Clara, 16, Broad street.
Constable, George, 39, Franklin.
Cope, Ahlum, 70, Conemaugh.
Costlow, Michael, 70, Locust and Union streets.
Costlow, Zita, 6, Woodvale.
Costlow, Juniata, 2½, Woodvale.
Costlow, Regina, 1, Woodvale.
Craig, William, 8, 314½ Walnut street.
Creed, Kate, 26, 200 Washington street.
Creed, Mary, 16, Washington street.
Crown, Thomas, 51, Conemaugh.
Crowthers, Samuel, 30, Cambria.
Crowthers, Mrs. Verna, 27, Cambria.
Culleton, George F., 1, Chestnut street.
Culleton, John F., 2, Chestnut street.
Cummings, Mrs., Somerset street.
Cummings, —, Somerset street.
Cunz, Mrs. Catherine, 37, Napoleon street.
Cunz, Edward, 12, Napoleon street.
Cunz, Gussie, 3, Napoleon street.
Curtin, Johanna.
Cush, Annie, 17, 112 Railroad street.
Cush, Thomas, 1½, 116 Railroad street.
Custer, Mrs. Emma J., 27, Bedford street.
Curry, Robert.
Darr, George E., 28, Millville.
Davis, Martha, 18, Woodvale.
Davis, Ada, 15, Woodvale.
Davis, Mrs. Ann, 60, Locust street.
Davis, Mrs. Mary, 54, Locust street.
Davis, Della, 22, Locust street.
Davis, Evan, 16, Locust street.
Davis, Reese, 13, Locust street.
Davis, Mrs. Mary D., 55, Millville.
Deible, Harry, Woodvale.
Deihl, Mrs. Mary, 40, Conemaugh.
Delaney, Charles, 18, 51 Conemaugh street.
Devlin, Melissa, 12, East Conemaugh.
Dick, Cornell, 17, Cor. Locust and John streets.
Dill, Robert, 26, Woodvale.
Dill, Mrs. Robert, 26, Woodvale.
Dill, William, 7, Woodvale.
Dill, Harry, 3 Woodvale.
Dinkel, Adam, 50, Conemaugh.
Dishong, Lizzie, 22, Union street.
Dluhos, Jacob, 3, Cambria.
Dluhos, Mary, 3 months, Cambria.
Dolny, Mike, Cambria.
Dorillia, Mrs., 30, Cambria.
Dorillia, —, Cambria.

Dorillia, —, Cambria.
Downs, Willie, Millville.
Dudzik, Mike, 21, Cambria.
Dudzik, Albert, 21, Cambria.
Early, Mrs. Ella, 59, Woodvale.
Eck, Ellen C., 6, Woodvale.
Eck, Edna Marie, 1½, Woodvale.
Eck, John B., 38, Conemaugh street.
Eck, Dora, 7, Conemaugh street.
Eck, Mabon, 2, Conemaugh street.
Edmonds, Nancy.
Edwards, Roger, 55, Millville.
Edinger, Annie, 19, Millville.
Elder, Mrs. Cyrus, 49, Walnut street.
Elder, Nannie M., 23, Walnut street.
Eldridge, Pennell, 39, Morrellville.
Eldridge, Mrs. Sarah T., 71, Woodvale.
Eldridge, Mrs. Sallie, 27, Woodvale.
Eldridge, Clara, 3, Woodvale.
Eldridge, Annie, 1, Woodvale.
Elsaesser, Constantine, 44, Railroad street.
Elsaesser, Mrs. Frances, 41, Railroad street.
Elsaesser, Charles, 13, Railroad street.
Elsaesser, Adolph, 11, Railroad street.
Elsaesser, Maggie, 10, Railroad street.
Elsaesser, Rose, 4, Railroad street.
Elsaesser, Mary, 1, Railroad street.
English, John.
Etchison, Samuel, 37, Hulbert House.
Evans, Evan B., 50, Woodvale.
Evans, Susannah, 16, Woodvale.
Evans, Mrs. Mary, 55, Main street.
Evans, Annie, 26, Millville.
Evans, Jennie, 13, Millville.
Evans, Susannah, 9, Millville.
Evans, Idris, 3, Millville.
Evans, Walter, 8, Vine street.
Evans, Albert, 12, Conemaugh.
Evans, Elizabeth.
Fairfax, Mrs. Susan, 94, Somerset street.
Fairfax, Mrs. G. W., 38, Somerset street.
Fedorizen, Miklosz, Cambria.
Fenlon, Patrick, 70, Conemaugh.
Fendra, E. H.
Fenn, John Fulton, 12, Locust street.
Fenn, Daisy, 10, Locust street.
Fenn, George Washington, 8, Locust street.
Fenn, Virginia, 5, Locust street.
Fenn, Esther, 1½, Locust street.
Fentiman, Edwin F., 19, Main street.
Fees, Frank, 23, Millville.
Fichtner, Carrie, Stonycreek street.
Fichtner, Annie, 21, Main street.
Fiddler, son of Jacob, Cambria.
Findlay, Mrs. Phœbe, 58, Woodvale.
Findlay, Robert B., 17, Conemaugh.
Fingerhute, Mary, 18, Conemaugh street.
Fingle, Mrs. Mary.
Fink, Samuel P., 54, Conemaugh street.
Fink, Mrs. Mary, 47, Conemaugh street.
Fisher, John, 60, Cambria.
Fisher, Johanna, 19, Cambria.
Fisher, Kate, 9, Cambria.
Fisher, Eddie, 7, Cambria.
Fisher, George, 3, Cambria.
Fisher, August.
Fisher, William.
Fitzgerald, Mrs. Catherine. 40, Millville.
Fitzpatrick, Eliza, 15 months, Cambria.
Fitzharris, Mary J., 16, Franklin street.
Fitzharris, Sarah A., 15, Franklin street.
Foling, August, Cambria.
Foster, Mrs. Margaret, 64, Woodvale.
Foster, Maggie, 29, Woodvale.

Frank, August, 26, Washington street.
Frank, Lena, 15, Washington street.
Fritz, Mrs. Matilda, 26, Horner street.
Fritz, Jane, 2, Horner street.
Fritz, Lily, 1, Horner street.
Gaffney, Mrs. Ellen, 26, Cambria.
Gaither, Willie, 15, South St.
Gardner, John.
Geczie, John, 47, Cambria.
Geczie, Veronica, 37, Cambria.
Geczie, Stephen, 8, Cambria.
Geczie, Annie, 4, Cambria.
Geczie, August, 2, Cambria.
Geczie, Belle, 3 months, Cambria.
Geddes, Mrs. George, 40, Woodvale.
Geisel, Julia, 9, Cambria.
Geisel, Rolla, 9, Cambria.
Geraldan, Mrs., 17, Conemaugh street.
Gillas, David, 66, Cambria.
Given, Cora, Millville.
Glass, James, 45, School Alley.
Golde, Harry, 5, Walnut street.
Gouchenour, Frank, 31, Conemaugh.
Grant, Mrs. Kate, 24, Cambria.
Grant, Bernard, 5, Cambria.
Grant, John, 3, Cambria.
Gray, Mrs. Frances, 36, Woodvale.
Gray, Gerald, 5, Woodvale.
Gray, Inez, 3, Woodvale.
Greenwood, Mary A., 8, Cambria.
Greitzer, George, 27, Cambria.
Greger, Ann.
Griffith, Mr.
Gromley, Mrs. Magdalena, 48, Mineral Point.
Gromley, Mary M., 16, Mineral Point.
Gromley, Daniel J., 13, Mineral Point.
Gromley, Emanuel L., 9, Mineral Point.
Gromley, Emma B., Mineral Point.
Hager, Mrs. Mary, 62, Washington street.
Hagerty, Mrs. Mary J., 36, School alley.
Hagerty, Kate, 12, School alley.
Hagerty, Stella, 8, School alley.
Haight, Annie, First Ward.
Haltie, Miss.
Haldiman, Hy, Woodvale.
Hammers, George.
Hammill, Mrs. Catherine, 70, Cambria.
Hamilton, Lou.
Hannan, Mamie, 22, Woodvale.
Hanekamp, Mrs. Louise, 28, Lincoln street.
Hanekamp, child.
Harrigan, Mary L., Millville.
Harris.
Hart, May, 9, Market street.
Harkey, William G.
Hess, Mrs.
Haugh, John, Conemaugh.
Hayes, Thos., 10, Cambria.
Hayes, Annie, 5, Cambria.
Hayes, Agnes Gertrude, 3 weeks, Cambria.
Heckman, Miss, 18, Cambria.
Heidenthal, Mrs. Mary, 38, Woodvale.
Heidenthal, Joseph, 14, Woodvale.
Heidenthal, Annie, 12, Woodvale.
Heidenthal, Phœbe, 10, Woodvale.
Heidenthal, Bertha, 6, Woodvale.
Heidenthal, Alfred, 2, Woodvale.
Heingard, Annie, 22, Woodvale.
Heine, Joseph, 1, Cambria.
Heine, Amelia, 8 months, Cambria.
Hellenberger, Miss E.
Hellreigle, Chas. J., 28, Woodvale.
Henahan, John, 40, Cambria.

Henahan, Mrs. Mary, 24, Cambria.
Henahan, Mary, 7, Cambria.
Henahan, Catherene, 4, Cambria.
Henahan, Frances, 1, Cambria.
Henderson, Thomas, South Fork.
Henderson, Robert, 6 months, Main street.
Henning, John.
Henning, Mary.
Hickey, Stephen, 9, Conemaugh.
Hicks, Miss Ella, Woodvale.
Himes, Charles C., Conemaugh street.
Himes, Mrs. C. C., Conemaugh street.

Himes,
Himes, } children.
Himes,

Hinchman, Franklin, 2, Woodvale.
Hirsch, Henry, 10, Cambria.
Hockenberger, Ann, Napoleon street.
Hoffman, Mrs. Mary, 69, Conemaugh.
Hoffman, Joseph, 10, Conemaugh.
Hoffman, Mary, 8, Conemaugh.
Hoffman, Peter, 78, Market street.
Hoffman, Frank C., 11, Market street.
Hoffman, Sehna, 3, Market street.
Hoffman, Lena, 19, Washington street.
Hoffman, George, 12, Washington street.
Hoffman, Crissie, 9, Washington street.
Hoffman, Albert, 4, Washington street.
Hoffman, Walter, 2, Washington street.
Hoffman, Stella, 6 months, Washington street.
Hoffman, Fred W., 42, Conemaugh.
Hoffman, Mrs. Jennie, 40, Conemaugh.
Hoffman, Lena, 19, Conemaugh.
Hoffman, Henry, 65, Conemaugh.
Hoffman, Mrs. Mary Ellen, 55, Conemaugh.
Hoffman, Stewart, 24, Conemaugh.
Hoffman, Mrs., Conemaugh.
Hoffman.
Hoffman.
Hoffman.
Hoopes, Mrs. Maria, 25, Woodvale.
Hoopes, Ernest, 5, Woodvale.
Hoopes, Allen C., 6 months, Woodvale.
Hopkins, Hannah, 40, Locust street.
Hopkins, Elizabeth, 4, Conemaugh.
Hopkins, Geo., 8, Conemaugh.
Hopp, Mary E., 7 months, Vine street.
Horner, Miss, Hulbert House.
Horner, Elwood, 15, Levergood street.
Hornick, Wm., 23, Conemaugh.
Hough, Mrs. Louisa, 48, Conemaugh.
Hough, Patrick, 5, Conemaugh.
Houghton, Mrs. Lizzie J., 24, Walnut street.
Howe, Mary E., Washington street.
Howe, Mrs. Nancy, 50, Bedford street.
Howe, Robert G., 8, Bedford street.
Howe, Mrs. W. J.,
Howells, Maggie, 15, Union street.
Howells, John, 25, Union street.
Howells, Wm., 4 days, Union street.
Hughes, Lizzie A., 1, Chestnut street.
Hurst, Mrs. Minnie, 60, Washington street.
Hurst, Emily, 10, Washington street.
Hammell, Margaret, 14, Washington street.
Illis, Daniel, Cambria.
James, John K., 8, Main street.
James, William, 10, Market street.
James, Mrs. John.
James, Benjamin, Third Ward.
Janosky, Mrs. Lena, 27, Market street.
Jenkins, John, 20, Upper Yoder.
Jenkins, Harvey, 6, Vine street.

Jenkins, Thomas, Third Ward.
Jenkins, Mrs. Thomas, Third Ward.
Jenkins, ——.
Johns, Mrs. Josephine, 32, Woodvale.
Johns, Richard, 14, Woodvale.
Johns, Silvie, 11, Woodvale.
Johns, Stephen, 5, Woodvale.
Johnson, Mrs. David, 40, Conemaugh.
Johnson, Geraldine, 17, Conemaugh.
Johnson, George, 17, Union street.
Johnson, William, 15, Union street.
Johnson, Gertrude, 13, Union street.
Johnson, Lottie, 11, Union street.
Johnson, Dollie, 7, Union street.
Johnson, Frederick, 4, Union street.
Johnson, Lulu, 3, Union street.
Johnson, Ellen, Hulbert House.
Jones, Mrs. Alice, 65, Millville.
Jones, Mrs. Rachael, 41, Main street.
Jones, Ella, 11, Main street.
Jones, Sarah, 8, Main street.
Jones, Abner, 6, Main street.
Jones, Ida, 3, Main street.
Jones, Thomas, 6, Conemaugh street.
Jones, Elmer, 2, Conemaugh street.
Jones, Mrs. Jennie, 50, Woodvale.
Jones, Williams, 17, Woodvale.
Jones, Amanda, 40, Woodvale.
Jones, Pearl, 9, Woodvale.
Jones, William, 4 months, Woodvale.
Jones, James, 19, Pearl street.
Jones, Charles, 16, Pearl street.
Jones, Emma, Second Ward.
Jones, Walter B., 7, Main street.
Jones, Mrs. Margaret, 65, Llewellyn street.
Jones, Rev. E. W., 56, Vine street.
Jones, Mrs. Rev. E. W., 55, Vine street.
Kane, Mrs. Lidia, 44, Union street.
Kane, Ellsworth, 18, Union street.
Kane, Laura, 15, Union street.
Kane, Willie, 12, Union street.
Kane, Dollie, 10, Union street.
Kane, Lester, 2, Union street.
Kane, Emma, 21, Prospect.
Kane, Mrs. Ann, 60, Cambria City.
Kast, Mrs. Charlotte, 43, Clinton street.
Kalor, Mrs. Philapena, 67, Conemaugh.
Kalor, Jamanes.
Kalor, Jane.
Kaylor.
Kaylor.
Kaylor.
Kaylor.
Keedy, Clay, 5, Millville.
Keelan, Mrs. Catherine, 55, Cambria.
Keelan, Daphne, 13, Cambria.
Keelan, Edward.
Keelan, Frank.
Keene, Mrs. Elizabeth, 60, Union street.
Keenan, Mrs. Jane, 26, Washington street.
Keiflein, Philamena, Conemaugh.
Keis, Mrs. Caroline, 24, Railroad street.
Keis, infant. Railroad street.
Kehoe, Thomas, 24, South Fork.
Kelly, Mary M., 30, Millville.
Kelly, Mary C., 1½, Millville.
Kelly, Maggie, 17 days, Millville.
Kelly, Mrs. Ann, 45, Cambria.
Kelly, John W., 24, Cambria.
Kidd, Mrs. Jenny, 35, Walnut street.
Kidd, Laura, 5, Walnut street.
Kilgore, Mrs. W. A., 48, Washington street.
Kilgore, Jessie, 15, Washington street.
Kilgore, Fred, 12, Washington street.

Kilgore, Alex., 9, Washington street.
Kimpel, Mrs. Christ, 43, Clinton street.
Kinder, Thomas, 40, Moxham.
King, Mrs. James, 48, Broad street.
King, Katie M., 24, Broad street.
King, James, 5, Broad street.
Kinney, Mrs. Margaret, 31, Washington street.
Kinney, ——, 4, Washington street.
Kinney, Agnes, 10, Washington street.
Kintz, Teresa, 24, Cambria.
Kinley, Jane, Bausman alley.
Kirkbride, Fannie, 11, Hager Block.
Kirkbride, infant, Hager Block.
Kirkwood, Finley, 19, Conemaugh.
Kirlin, Mrs. Thomas, 32, Conemaugh street.
Kirlin, Willie, 2, Conemaugh street.
Knable, John.
Knable, Leonard.
Knox, Thomas, 54, Somerset.
Keohler, Mrs. Philomen, 87, Conemaugh.
Keohler, Wm., 16, Conemaugh.
Kraft, Mrs. Maggie, 37, Walnut street.
Kraft, Herman, 12, Walnut street.
Kraft, Frederick, 10, Walnut street.
Krieger, Katie.
Kunkle, Katie, 19, Washington street.
Lambreski, Willie, 2, Cambria.
Lavelle, Miss Mary, 31, Broad street.
Lavelle, Kate, 24, Broad street.
Lavelle, Sallie, 18, Broad street.
Lavelle, Mrs. Mary, 58, Broad street.
Lavelle, John F., 8, Conemaugh street.
Lavelle, Edgar R., 4, Conemaugh street.
Lavelle, Frances A., 6, Conemaugh street.
Laystrom, Mrs., 30, Union street.
Layton, infant, Broad street.
Lewis, Mrs. Lizzie, 28, Lewis alley.
Lichtenberg.
Lichtenberg.
Lichtenberg.
Lightner, James, 23, Cambria.
Lightner, Mrs. Mary, 21, Cambria.
Lightner, Eddie, 1, Cambria.
Lohr, Julia, 17, Bedford street.
Lonaenstein, Mrs. Ida, 27, Franklin.
Ludwig, Charles E., 30, Railroad street.
Luther, Michael, 40, Cambria.
Madden, Willie, 12, Cambria.
Maloy, Mrs. Ann, 35, Millville.
Maloy, Jane, Hulbert House.
Marks, William.
Martin, Edward, 48, River avenue.
Martin, Mrs. Catherine, 40, Millville.
Martin, Mary, 18, Millville.
Martin, Ann, 7, Millville.
Martin, Celia, 7, Millville.
Masters, Margaret.
Masterton, Miss.
Mayhew, Annie, 12, Woodvale.
Mayhew, Earnest, 9, Woodvale.
Mayhew, Harry, 6, Woodvale.
McAteer, Mrs. Jane, 38, Cambria.
McAneny, Sarah, 7, Cambria.
McAley, P.
McCann, Mrs. John, 30, Railroad street.
McCann, John, 31, Railroad street.
McCann, infant, Railroad street.
McClarren, Mary, 13, Cambria.
McClarren, Philip, 1½, Cambria.
McConaghy, Harry M., 6, Main street.
McConaghy, Frank A., 2, Main street.
McCoy, Mr., Railroad street.
McGrew, Oscar, Conemaugh.

McGuire, Constantine, 48, Woodvale.
McGuire, Ann, 19, Woodvale.
McGuire, Christian, 17, Woodvale.
McHugh, Kate, 19, Cambria.
McKeever, Mrs. Mary.
McKim, Mrs. Polly, 65, East Conemaugh.
McMeans, William, 33, Conemaugh street.
McPike, Rosie, 4, Cambria.
McWilliams, Susie, 13, Pittsburgh.
Melczer, Robert, 35, Cambria.
Melczer, Mrs. Johanna, 30, Cambria.
Melczer, Albert, three weeks, Cambria.
Melczer, Mary, 4, Cambria.
Melczer, John, 2, Cambria.
Merle, Mrs. George, Washington street.
Merle, Mrs. Ida, 29, Washington street.
Merle, Nettie, 5, Washington street.
Merle, Elmer, 2, Washington street.
Meredith, Mr. (probably duplicate).
Meyers, Joseph, 70, Cambria.
Meyers, Lizzie, 11, Millville.
Meyers, Annie, 9, Millville.
Meyers, Stella, 7, Millville.
Meyers, Charlie, 5, Millville.
Meyers, Philip, 1, Millville.
Michalitch, Mrs. Mary, 38, Cambria.
Michalitch, Martin, 6, Cambria.
Michalitch, Mary, 3, Cambria.
Michalitch, John, 1, Cambria.
Miller, Lizzie, 11, Woodvale.
Miller, John, 1, Cambria.
Miller, Mrs. Sophia, 45, Cambria.
Miller, John, 8, Cambria.
Miller, Mary, 12, Horner street.
Monteverde, Mrs. Maria, 40, Washington street.
Monteverde, Joseph, 5, Washington street.
Monteverde, Eleanora, 1½, Washington street.
Moore, Melda, 20, Main street.
Moreland, Mrs. Margaret, 48, Quarry street.
Morgan, Gertie, 11, Millville.
Morgan, Mrs. Mary R., 66, Conemaugh street.
Morgan, Miss, Conemaugh.
Moser, Heinrich, Cambria.
Moschgat, Amelia, 22, Bedford street.
Mullen, Mrs. Mary, 65, Conemaugh street.
Mullen, Mrs. Margaret, 47, Prospect.
Mumma, Mrs. Eliza, 26, Washington street.
Murphy, Mrs. Kate H., 48, Park Place.
Murphy, Mrs. Maggie, 34, Brunswick Hotel.
Murphy, John, 10, Brunswick Hotel.
Murphy, Clara, 8, Brunswick Hotel.
Murphy, Genevieve, 6, Brunswick Hotel.
Murphy, Martin F., 4, Brunswick Hotel.
Murphy, Maggie, 2, Brunswick Hotel.
Murr, Stella, 16, Washington street.
Murr, Frederick, 11, Washington street.
Murr, Nellie, 6, Washington street.
Murr, Frida, 3 months, Washington street.
Nadi, Frank.
Nainbaugh, Henry.
Nayuska, Mrs. Hannah, 65, Market street.
New, Frank.
Newell, August.
Newman, Banheim, 68, Washington street.
Nich, Peter, 30, Cambria.
Nich, William, 2, Cambria.
Nich, Lena, 23, Cambria.
Nich (infant), Cambria.
Nix, Frank, Cambria.
Nixon, Fannie, 5, Woodvale.
Neice, Conrad.

O'Brien, Mrs. Sarah, 60, Millville.
O'Brien, Mrs. Ellen, 31, Millville.
O'Brien, Mrs. Catherine, 55, Millville.
O'Callahan, James, 70, Millville.
O'Callahan, Mrs. Bridget, 68, Millville.
O'Callahan, Miss Ella, 25, Millville.
O'Connell, Edward, Cambria.
O'Donnell, Mrs. Julia, 26, Washington street.
O'Donnell, John, 2, Washington street.
Ogle, Mrs. Hettie M., 52, Washington street.
Ogle, Minnie T., 32, Washington street.
Oberlander, Robert, 35, Locust street.
Oberlander, Mrs. Robert, 30, Locust street.
Oberlander, Mary, 2, Locust street.
O'Lily, Catherine, 20, Cambria.
O'Neill, James, 2, Cambria.
Oswald, Eulaliah, 9, Third Ward.
Osterman, Mrs. Victoria, 31, Cambria.
Osterman, Conrad, 4, Cambria.
Osterman, Joseph Jr., 6, Cambria.
Osterman, Mary Ann, 1½, Cambria.
O'Shea, Mary, Second Ward.
Owens, Mrs. Mary, 62, Market street.
Owens, John, 12, Conemaugh street.
Owens, Amelia, 6, Conemaugh street.
Owens, Willie, 4, Conemaugh street.
Owens, Mrs. Elizabeth, 37.
Pfeifer, Charles, 30, Woodvale.
Pfeifer, Ella, 21, Woodvale.
Pheng, John, Conemaugh.
Phillips, Mary, 16, Union street.
Phillips, Grace, 12, Union street.
Phillips, John J., 14, Market street.
Phillips, David, 12, Market street.
Phillips, Richard, 10, Market street.
Phillips, Mary, 8, Market street.
Phillips, Evan, 6, Market street.
Pipple, Mrs., Fourth Ward.
Plummer, Alvin.
Pollocks, Louis, 19, Cambria.
Polk, John.
Potts, Mrs. Mary, 29, Market street.
Powell, Mrs. Reese, 74, Main street.
Pratt, ——, Cambria.
Pratt, ——, Cambria.
Pritchard, Mrs. Henry, 48, Market street.
Pritchard, Howell, 9, Market street.
Pritchard, Alice, 5, Market street.
Pritchard, Rachael, 3, Market street.
Price, Mrs. Abe, 29, Millville.
Progner, Samuel, 28, Conemaugh.
Prosser, Mrs. David, 68, Union street.
Pukey, Julius, 23, Cambria.
Pukey, Matilda, 1, Cambria.
Raab, Mollie, 18, Clinton street.
Raab, Bertha, 13, Clinton street.
Raab, Katie, 3, Clinton street.
Raab, Mrs. Minnie, 24, Washington street.
Rawn, Mrs. Henrietta, 78, Conemaugh.
Ream, Mrs. Mary, 34, Woodvale.
Ream, Joseph, 10, Woodvale.
Ream, Effie May, 6, Woodvale.
Ream, Cora, 1, Woodvale.
Ream, Frederick E., 23, Third Ward.
Ream, Amelia, 20, Third Ward.
Reamus, Gussie, 17, Woodvale.
Recke, Mrs. Alex., 29, Washington street.
Reed, Charles.
Reese, Susie, 14, Millville.
Reese, Sarah, Woodvale.
Reese, Mrs., 70.
Reidel, Mrs. Teresa, 56, Conemaugh.
Reilly, Timothy, 27, Millville.

Ressler, John.
Reynolds, Mrs. Elizabeth, 40, Woodvale.
Reynolds, Idella, 14, Woodvale.
Reynolds, Columbia, Conemaugh.
Rhodes, Frank, 2, Somerset street.
Rich, Harry, 16, Stonycreek street.
Richards, Mrs. Margaret, 40, Union street.
Riffle, Mary C., Cambria.
Riley, Mrs. Bridget, 40, Cambria.
Riley, Annie, 8, Cambria.
Riley, Katie, 6, Cambria.
Ripple, Emma, 24, Bedford street.
Ritter, Katie, 20, Cambria.
Ritter, Sophia, 12, Cambria.
Rodgers, Mary, Hulbert House.
Rodgers, Mary G., 19, Woodvale.
Rodgers, Mrs. Mary, Water street.
Roland, Lizzie, 5 months, Conemaugh.
Roose, John, 31, Haynes street.
Rosenfelt, ——.
Rosenfelt, ——.
Rosenfelt, ——.
Rosenfelt, ——.
Rosenfelt, ——.
Rosensteel, Matilda V., 19, Woodvale.
Roth, Albert, 8, Cambria.
Roth, Mary, 6, Cambria.
Roth, Sebastian, First Ward.
Ruth, John.
Rowland, Mrs. E. J., 64, Market street.
Ryan, Sadie, 16, Washington street.
Ryan, Gertie, 3, Washington street.
Ryan, Mary, Third Ward.
Sagerson, Mrs., 96, Millville.
Salenty, E.
Sample, Mrs. Catherine, 63, East Conemaugh.
Sarlous, Grace, 16, Cambria.
Savage, Mrs. Bridget, 76, Woodvale.
Saley, Joseph, 50, Millville.
Shaffer, Mrs. Mary, 43, Cambria.
Shaffer, Carl, 19, Cambria.
Schanvisky, August, 10, Cambria.
Sherer, Mrs. Kate, 49, Conemaugh.
Sherer, Emma, 24, Conemaugh.
Sherer, Mary, 11, Conemaugh.
Schiffhauer, Frances, 19, Washington street.
Schittenhelm, Wilmena.
Schmitt, William J., 7, Cambria.
Schmitt, Mrs. Augustina, 38, Cambria.
Schmitt, August, 8, Cambria.
Schmitt, Anton, 2, Cambria.
Schmitt, Annie, 1, Cambria.
Schmitz, Ferdinand, Cambria.
Schmitz, Gabriel, 50, Conemaugh.
Schmidt, John L., Cambria.
Schonhardt, Victoria, 56, Conemaugh.
Schultz, Mrs. William, Clinton street.
Schultz, Clinton street.
Schultz, Clinton street.
Schultz, Clinton street.
Schultz, Clinton street.
Schultz, Joseph, First Ward.
Schweitzer, William, Conemaugh.
Schweitzer, Catherine E., Conemaugh.
Schurtz, Peter, 38, Conemaugh.
Seibert, Mrs. Elizabeth, 56, Woodvale.
Schaffer, Howard, 21, South Fork.
Shea, Mrs. Mary, 30, Locust street.
Sheldon, H.
Sherman, Mrs. Ann, 35, Market street.
Shinkey, Mrs., Second Ward.
Shorper, Jacob.
Shorper, Jacob, Jr.

Silverman, Moses, Second Ward.
Seigmund, Mrs. Matilda, 52, Woodvale.
Seigmund, Mrs. Carolina, 28, Woodvale.
Seigmund, John, 20, Woodvale.
Singer, Mrs. E. H., Unionport, Ohio.
Siroczki, Mrs. Mary, 30, Cambria.
Siroczki, Mary, 7, Cambria.
Siroczki, Annie, 4, Cambria.
Siroczki, Lizzie, 2, Cambria.
Skiba, Annie, 6, Cambria.
Skiba, Sophia, 1½, Cambria.
Smith, Harry, 5, Woodvale.
Smith, Hattie, 4, Woodvale.
Smith, infant, 3, Woodvale.
Smith, Alice J., 2, Woodvale.
Smith, Clarence, 6 months, Woodvale.
Smith, George A., 38, Pearl street.
Smith, Mrs. Jennie, 36, Pearl street.
Smith, Charles, 7, Pearl street.
Smith, Alum, 4, Pearl street.
Smith, Effie, 9 months, Pearl street.
Smith, Mrs. Mary, 21, Cambria.
Smith, Mollie, 22, Cambria.
Smith, Mrs. Ann, 55, Cambria.
Smith, Francis, 3, Cambria.
Smith, Charles, 1, Cambria.
Smith, John M., 38, Millville.
Smith, William, 9, Millville.
Smith, Mrs. Mary, Third Ward.
Smith, William, Third Ward.
Smith, Esther, Third Ward.
Smith, Charles, First Ward.
Smith, Richard, First Ward.
Smith, Frank, First Ward.
Snyder, Polly, 14, Woodvale.
Snyder, William, 8, Woodvale.
Snyder, Annie, 6, Woodvale.
Snyder, John, 3, Woodvale.
Snyder, Patrick V., 5 months, Woodvale.
Snyder, Hollis, Woodvale.
Snyder, Mary.
Snyder, Annie.
Snyder, John.
Snyder, Mary E.
Snyder, Harrison V.
Speers, Mrs. L. E.
Spenger, Mrs. Catherine, 56, Stonycreek
    street.
Spenger, Edward, 16, Stonycreek street.
Spoller, Mrs.
Spoller, Lee.
Stansfield, James C., 30, Woodvale.
Stansfield, Mrs. J. C., 25, Woodvale.
Stansfield, Ralph, 9 weeks, Woodvale.
Steckman, Fred, 42, Cambria.
Stewart, Watson, 60, Pearl street.
Stewart, Mrs., 70, Walnut street.
Stews, Louis, Walnut street.
Stinely, Annie, 4, Cambria.
Stinely, infant, 4 months, Cambria.
Stork, Casper, 43, Walnut street.
Stork, Mary, 38, Walnut street.
Stork, John, 20, Walnut street.
Stork, Lizzie, 14, Walnut street.
Strauss, Charles S., Conemaugh.
Strayer, Katie, 22, Market street.
Strayer, Bertha, 14, Market street.
Stroup, Henry, Conemaugh.
Stufft, Vera, 10, Woodvale.
Stufft, Earl B., 8, Woodvale.
Stufft, Lula B., 6, Woodvale.
Stufft, Elda M., 3, Woodvale.
Stufft, infant, four months, Woodvale.
Suder, Lizzie, 9, Millville.

Suder, James, 5, Millville.
Sullivan, Mrs. Catherine, 55, Millville.
Swank, Leroy, 4, Main street.
Swank, Miss, Morris street.
Sweitzer, William, 35, Morrellville.
Temple, Leroy.
Thoburn, John, 40, Millville.
Thoburn, Mrs. Flora, 36, Millville.
Thoburn, John, Jr., 10, Millville.
Thoburn, Harry, 1, Millville.
Thomas, Tydvil, 19, Millville.
Thomas, Mrs. Annie E., 56, Napoleon
    street.
Thomas, Mrs. Ann, 41, Woodvale.
Thomas, Albert E., 17, Woodvale.
Thomas, Vivian D., 15, Woodvale.
Thomas, James Roy.
Thomas, Sylvester.
Thomasberger, Fannie, 42, Conemaugh.
Thomasberger, Nellie, 13, Conemaugh.
Thomasberger, Charles, 11, Conemaugh.
Thurin, Levi.
Totas, Jacob, Cambria.
Totas, Sophia, Cambria.
Totas, Michael, Cambria.
Totas, Wavreck, Cambria.
Trefts, William S.
Tross, W. J. Sr., 43, Woodvale.
Tross, Katie, 19, Woodvale.
Tross, William, 17, Woodvale.
Tross, Conrad, 16, Woodvale.
Tross, Charles, 13, Woodvale.
Tross, George, 9, Woodvale.
Tross, Louis, 7, Woodvale.
Tross, Edward, 6, Woodvale.
Tucker, Mrs. Margaret N., 45, Woodvale.
Tucker, Lillian G., 18, Woodvale.
Tucker, Mabel, 6, Woodvale.
Tynan, Michael J., 49, Conemaugh.
Tynan, Mrs. M. J., 47, Conemaugh.
Unverzagt, Lulu, 23, Washington street.
Vallance, David, 55, Conemaugh street.
Vallance, Mrs. Sarah, 66, Conemaugh street.
Vallance, Annie, 21, Conemaugh street.
Valentine, Mrs. Carrie, Market street.
Valentine, Alexander L., 14, Market street.
Valentine, Annie May, 11, Market street.
Valentine, Burt, 7, Market street.
Valentine, Howard, 4, Market street.
Valentine, Ruth, 1½, Market street.
Varner, Viola, 12, Cambria.
Varner, Sarah, 10, Cambria.
Varner, Ida, 7, Cambria.
Varner, Ella, 5, Cambria.
Varner, infant, six weeks, Cambria.
Veith, Mrs. Carrie, 52, Stonycreek street.
Veith, Emma, 14, Stonycreek street.
Voeghtly, Mrs.
Von Alt, Henry, Clinton street.
Wagnor, Mrs. Henry, Cambria.
Wagnor, Frank, Cambria.
Wagnor, John, Cambria.
Walker, Conrad, 27, Clinton street.
Walker, Ida J., 22, Conemaugh.
Walser, Mrs. Ann, Alum Bank, Pa.
Ward, Ella, Cambria.
Warren, Edward, 28, Millville.
Waters, Thomas J., 15, Conemaugh.
Watkins, Mary J., 22, Washington street.
Wearn, Mrs. Priscilla, 66, Walnut street.
Wearn, Richard, 30, King street.
Wearn, Mrs. Ella, 27, King street.
Wearn, Myrtle, 3, King street.
Weaver, Joseph H., 19, Woodvale.

Weaver, Margaret J., Second Ward.
Webber, Christian, 31, Woodvale.
Wehelco, John, Cambria.
Wehn, Casper, 80, Clinton street.
Weinzarl, Annie, 13, Cambria.
Weinzarl, Martha, 11, Cambria.
Weinzarl, Sarah, 7, Cambria.
Weinzarl, Mollie, 5, Cambria.
Weinzarl, John, 3, Cambria.
Weinzarl, George, 4 months, Cambria.
Weisc, Rosa, 10, Cambria.
White, Annie, 23, Market street.
White, Raymond, 4, Youngstown, O.
Wickersham, Richard G., 26, Woodvale.
Wilson, Mrs. Lavina, 38, East Conemaugh.
Wilson, James, 33, Mineral Point.
Wilson, Henry, 58, Millville.
Wilson, Mr., Cambria.
Wiseman, Charles, 26, Conemaugh.
Wiseman, Emma, 4, Conemaugh.
Wiseman, August, 2, Conemaugh.
Witz, Sarah, Third Ward.
Wolf, Anthony, 24, Cambria.
Wolf, Albert, 1½, Cambria.

Wolford, Andrew, Conemaugh.
Wolford, Conemaugh.
Wolford, Conemaugh.
Wolford, Conemaugh.
Woren, Richard, Walnut street.
Woren, Mrs. Richard, Walnut street.
Woren, Willie, 6, Walnut street.
Woren, ——, child, Walnut street.
Woren, ——, child, Walnut street.
Woren, Mrs. Priscilla, 60, Walnut street.
Woren, Miss, 24, Walnut street.
Woren, Mrs. Thomas, Walnut street.
Yocum, Samuel, Third Ward.
Yost, Laura, 18, Grant street.
Yost, Lottie, Jackson street.
Young, Mamie, 12, Broad street.
Young, Katie, 10, Broad street.
Youst, Mr.
Youst, Eddie.
Zellar, Rose.
Zern, Miss.
Zimmerman, Milton, 19, Locust street.
Zimmerman, Morgan, 11, Young's alley.
Zimmerman, Owen N.

Bibliography

Alexander, Edwin P., *The Pennsylvania Railroad, A Pictorial History*. W. W. Norton & Co., New York, 1947.

Amory, Cleveland, *The Last Resorts*. Harper & Brothers, New York, 1952.

Andrews, J. Cutler, *Pittsburgh's Post-Gazette*. Chapman & Grimes, Boston, 1936.

Barton, William E., *The Life of Clara Barton, Founder of the American Red Cross*. Houghton Mifflin Co., Boston and New York, 1922.

Beale, The Reverend David J., *Through the Johnstown Flood*. Hubbard Brothers, Philadelphia, Boston, 1890.

Biographical Directory of the American Congress, 1774–1961. U.S. Government Printing Office, Washington, D.C., 1961.

Biographical and Portrait Cyclopedia of Cambria County. Philadelphia, 1896.

Bishop, Philip W., *The Beginnings of Cheap Steel*. Smithsonian Institution, Washington, D.C., 1959.

Bridge, James Howard, *The Inside History of the Carnegie Steel Company*. The Aldine Book Co., New York, 1903.

Bruce, Robert V., *1877: Year of Violence*. The New Bobbs-Merrill Co., Indianapolis-New York, 1959.

Burgess, George H., and Kennedy, Miles C., *Centennial History of the Pennsylvania Railroad Company 1846–1946*. Pennsylvania Railroad Co., Philadelphia, 1949.

Carnegie, Andrew, *Autobiography*. Houghton Mifflin Co., Boston and New York, 1920.

Carson, Herbert N., *The Romance of Steel, The Story of a Thousand Millionaires*. A. S. Barnes, New York, 1907.

Century Cyclopedia of History and Biography of Pennsylvania. Century Publishing Co., Chicago, 1904.

Chambers, Julius, *Lovers Four and Maidens Five, a Story of the Allegheny Mountains*. Porter & Coates, Philadelphia, 1886.

Chapman, The Reverend H. L., *Memoirs of an Itinerant*. Privately published, no dates or place of publication listed.

Clark, C. B., *C. B. Clark's Johnstown Directory and Citizens Register*. Altoona, 1889.

Clark, Victor S., *History of Manufacturers in U.S.*, Vol. II, *1860–1893*. Peter Smith (with permission Carnegie Institute of Washington), New York, 1949.

Connelly, Frank, and Jenks, George C., *Official Hisory of the Johntown Flood*. Journalist Publishing Co., Pittsburgh, 1889.

Diary of Isadore Lilly. Ebensburg, 1889.

Dictionary of American Biography. Charles Scribner's Sons, 1937.

Dieck, Herman, *The Johnstown Flood*, Philadelphia, 1889.

Epler, Percy H., *The Life of Clara Barton*. The Macmillan Co., New York, 1915.

Ferris, George T., *The Complete History of the Johnstown and Conemaugh Valley Flood*. H. S. Goodspeed & Co., New York, 1889.

Field, The Reverend C. N., *After the Flood*. Philadelphia, 1889.

Fitch, John A., *The Steel Workers*. Russell Sage Foundation, New York, 1911.

General Alumni Catalogue of University of Pennsylvania, 1922.

Grand-View Cemetery, An Historical Sketch. Johnstown, 1931.

Hall, John W., *Fact Sheet of the Flood of 1889*. (Compiled for the 75th anniversary of the disaster.)

Harvey, George, *Henry Clay Frick*. Charles Scribner's Sons, New York, 1928.

Heiser, Victor, M.D., *An American Doctor's Odyssey*. W. W. Norton & Co., New York, 1936.

Hendrick, Burton J., *Life of Andrew Carnegie*. Doubleday, Doran & Co., New York, 1932.

Hendrick, Burton J., and Henderson, Daniel, *Louise Whitfield Carnegie; The Life of Mrs. Andrew Carnegie.* Hastings House, New York, 1950.

Historical Sketch and Manual of Shady Side Presbyterian Church. Pittsburgh, 1892.

Holbrook, Stewart H., *The Age of the Moguls.* Doubleday, New York, 1954.

Illustrated Historical Atlas Combination of Cambria County. Philadelphia, 1890.

Interviews with flood survivors filmed by WJAC-TV, Johnstown, 1964.

Jackson, R. M. S., M.D., *The Mountain.* J. B. Lippincott & Co., 1860.

Johnson, Tom L., *My Story.* B. W. Huebsch, New York, 1911.

Johnson, Willis Fletcher, *History of the Johnstown Flood.* Edgewood Publishing Co., Philadelphia, 1889.

Johnstown Borough Council Minute Book, 1889.

Josephson, Matthew, *The Robber Barons.* Harcourt, Brace and Co., New York, 1934.

Langford, Gerald, *The Richard Harding Davis Years.* Holt, Rinehart and Winston, New York, 1961.

Lorant, Stefan, *Pittsburgh, The Story of an American City.* Doubleday, New York, 1964.

McLaurin, J. J., *The Story of Johnstown.* James M. Place, Harrisburg, 1890.

Matthews, Donald H., Jr., *Presbyterians in the Upper Conemaugh Valley.* Johnstown, 1957.

Monthly Weather Review. United States Signal Service, May 1889.

Morison, Elting E., *Men, Machines, and Modern Times.* M.I.T. Press, Cambridge, 1966.

National Encyclopedia of American Biography.

O'Connor, Harvey, *Mellon's Millions.* John Day Co., New York, 1933.

O'Connor, Richard, *Johnstown the Day the Dam Broke.* J. B. Lippincott Co., Philadelphia, 1957.

Operations of the Board of Health in Consequence of the Floods at Johnstown. Harrisburg, 1891.

Orvis, Charles F., *Fishing With the Fly.* Manchester, Vermont, 1883.

Pittsburgh and Allegheny Illustrated Review. Pittsburgh, 1889.

Pittsburgh Directory 1887–88.

Reade, Charles, *Put Yourself in His Place.* 1870.

Recollections of flood survivors recorded on questionnaires put out by the Greater Johnstown Chamber of Commerce in commemoration

of the 75th anniversary of the disaster; on file at the Cambria
County Historical Society, Ebensburg.

Record Books, Court of Common Pleas. Pittsburgh.

Report of Citizens Relief Committee of Pittsburgh. Pittsburgh, 1890.

*Report on the Dedication of the Monument to the Unknown Dead
Who Perished in the Flood at Johnstown, May 31, 1889.*

Report of the Secretary of the Flood Relief Commission. Harrisburg,
1890.

Sesquicentennial of Cambria County. Cambria County Historical Society, Ebensburg, 1954.

Shank, William H., *The Amazing Pennsylvania Canals.* The Historical
Society of York County, Pennsylvania, 1965.

Shappee, Nathan D., *A History of Johnstown and the Great Flood of
1889: A Study of Disaster and Rehabilitation.* Thesis filed at the
University of Pittsburgh, 1940.

Sipes, William B., *The Pennsylvania Railroad.* Pennsylvania Railroad
Co., Philadelphia, 1875.

Slattery, Gertrude Quinn, *Johnstown and Its Flood.* Wilkes-Barre,
1936.

Storey, Henry Wilson, *History of Cambria County, Pennsylvania.*
Lewis Publishing Co., New York, 1907.

Strayer, Harold H., and London, Irving L., *A Photographic Story of
the 1889 Johnstown Flood.* Johnstown, 1964.

Swank, James M., *Cambria County Pioneers.* Allen, Lane & Scott, Philadelphia, 1910.

————, *History of the Manufacture of Iron in All Ages.* American
Iron and Steel Association, Philadelphia, 1892.

Taft, Pauline Dakin, *The Happy Valley (The Elegant Eighties in Upstate New York).* Photographer, Leonard Dakin. Syracuse University Press, Syracuse, New York, 1965.

*Testimony Taken by the Pennsylvania Railroad Following the Johns-
town Flood of 1889.* Two copies in possession of Irving London,
Johnstown, and the author.

*Toasts and Responses at the Banquet Given by the Chamber of Com-
merce of Pittsburgh*, May 27, 1892, at the Duquesne Club.

Transactions American Society of Civil Engineers. June 1891.

Van Rensselaer, M. G., *In the Heart of the Alleghenies.* Allen, Lane &
Scott, Philadelphia, 1885.

*Veteran Employes Association of the Pittsburgh Division of the Penn-
sylvania Railroad.* Pittsburgh, 1902.

Victims of the Johnstown Flood. Johnstown, 1890.

Walker, James H., *The Johnstown Horror*. L. P. Miller & Co., Chicago, 1889.

Wallace, Paul A. W., *Pennsylvania, Seed of a Nation*. Harper & Row, New York, 1962.

Watkins, J. Elfreth, *History of the Pennsylvania Railroad Company, 1846–96*. Withheld from publication, bound proofs.

Weil, The Reverend Carl, *The Rider of Johnstown*. Pittsburgh, 1892.

Wilson, William Bender, *History of the Pennsylvania Railroad Company*. H. J. Coates & Co., Philadelphia, 1899.

Winkler, J. K., *Incredible Carnegie*. Vanguard Press, New York, 1931.

World's Charity to the Conemaugh Valley Sufferers and Who Received It, The. Harry M. Benshoff, Publisher, Johnstown, 1890.

MAGAZINES

Engineering News
Forest and Stream
Harper's New Monthly Magazine, August 1883
Harper's Weekly
Iron Age, The
Le Monde Illustré, June 29, 1889
Leslie's Weekly
North American Review, August 1889
Pennsylvania Magazine of History and Biography (which contains, in its July and October issues of 1933, much of a semi-official but incomplete report on the disaster, written by the noted historian, John B. McMaster)

NEWSPAPERS

Boston Daily Globe
Boston Morning Journal
Boston Post
Cambria Freeman
Chicago Herald
Cincinnati Enquirer
Greensburg Tribune and Herald
Johnstown Daily Democrat
Johnstown Daily Tribune

Johnstown Tribune-Democrat
Johnstown Weekly Democrat
London Times
Mansfield (Mass.) News
New York Daily Graphic
New York Herald
New York Sun
New York Times
New York Tribune
New York World
Philadelphia Press
Pittsburgh Commercial Dispatch
Pittsburgh Post
Pittsburgh Post-Gazette
Pittsburgh Press
Rocky Mountain News (Denver)
Utica Saturday Globe

Index

Also available by
DAVID McCULLOUGH

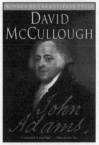

0-7432-2313-6

"**A masterwork** of storytelling."
—Walter Isaacson, *Time*

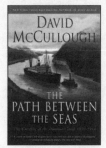

0-671-24409-4

"Dramatic, accurate…and
altogether gripping."
—*The Washington Star*

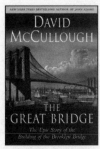

0-671-45711-X

"**One of the best books
I have read in years**."
—Robert Kirsch,
The Los Angeles Times

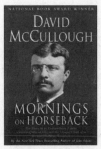

0-671-44754-8

"A **beautifully told** story,
filled with fresh detail."
—*The New York Times
Book Review*

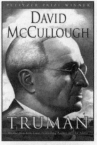

0-671-86920-5

"Warm, affectionate and
thoroughly captivating."
—*The New York Times
Book Review*

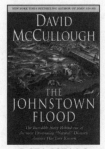

0-671-20714-8

"A suburb job, scholarly, yet **vivid**,
balanced yet **incisive**."
—*The New York Times*

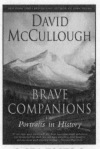

0-671-79276-8

"**All his subjects
come alive**."
—*The Dallas Morning News*

**SIMON & SCHUSTER
PAPERBACKS**
A VIACOM COMPANY